BEYOND RECYCLING

Beyond Recycling critically explores unasked questions around recycling and its prominent position in contemporary thinking about sustainability. It examines and challenges assumptions about why we appear to have so wholeheartedly committed to recycling as a cultural project.

Recycling has become a commonplace notion and widespread practice. Yet its social, cultural and even environmental value has not been considered carefully enough. This book considers recycling as a contemporary cultural idea related to – but not wholly defined by – our response to material waste. It seeks to reclaim recycling from the environmentalists and waste management specialists, to explore the role it plays in wider contemporary discourse. As we become increasingly satiated, and in many cases sickened, by the excesses of modern consumerism, we are rethinking our relationship with the physical stuff that fills our lives. Dissatisfied with empty materialism, we seek new ways to reuse our material culture. Recycling, turning something considered to be waste into something with renewed value, is our primary collective response to the problems arising from consumption; and it is ripe for critical examination.

Beyond Recycling is a fascinating read for conscious consumers and students of the creative arts, design, cultural studies, sustainability and environmental studies.

Paul Micklethwaite is Senior Lecturer at Kingston University, UK.

BEYOND RECYCLING

Paul Micklethwaite

Routledge
Taylor & Francis Group
LONDON AND NEW YORK

First published 2021
by Routledge
2 Park Square, Milton Park, Abingdon, Oxon OX14 4RN

and by Routledge
605 Third Avenue, New York, NY 10158

Routledge is an imprint of the Taylor & Francis Group, an informa business

British Library Cataloguing-in-Publication Data
A catalogue record for this book is available from the British Library

Library of Congress Cataloging-in-Publication Data
Names: Micklethwaite, Paul, author.
Title: Beyond recycling / Paul Micklethwaite.
Description: Milton Park, Abingdon, Oxon; New York, NY: Routledge, 2021. |
Includes bibliographical references and index.
Identifiers: LCCN 2020049712 (print) | LCCN 2020049713 (ebook)
Subjects: LCSH: Recycling industry. | Sustainability.
Classification: LCC HD9975.A2 M493 2021 (print) |
LCC HD9975.A2 (ebook) | DDC 363.72/82–dc23
LC record available at https://lccn.loc.gov/2020049712
LC ebook record available at https://lccn.loc.gov/2020049713

ISBN: 978-0-367-90388-6 (hbk)
ISBN: 978-0-367-90387-9 (pbk)
ISBN: 978-1-003-02408-8 (ebk)

Typeset in Futura
by Newgen Publishing UK

Beyond Recycling is dedicated to my mum, Cathy

CONTENTS

FIGURES

PREFACE

This book is driven by frustration at the way in which recycling dominates conversations about sustainability. Recycling is a dangerous idea if it prevents us thinking more deeply about how we can respond meaningfully to the global crisis of unsustainability.

The book's cover features a human skeleton doing the recycling. He holds two bags: a small transparent one containing recyclables and a larger black one containing non-recyclable waste, which is spilling its contents. The skeleton stands proudly doing his duty, offering with open arms his carefully sorted materials to a mysterious recycling system, which he hopes knows just what to do with them. His pose is also sacrificial; he is stuck in a loop from which there is no escape, perpetually consuming and expelling resources. The image forces us to consider: are we, in ignoring its limitations, recycling to our deaths? Finally, the skeleton's pose expresses the need for balance in our search for climate justice. We cannot address unsustainability by recycling alone but must consider every aspect of how we currently seek to meet our needs as humans on the Earth.

ACKNOWLEDGEMENTS

Ben Hayes initially commissioned me to write a book that was critical of recycling, working with me to shape the first proposal, and encouraging the broad scope of the project; I hope this final published version lives up to that early ambition. Many anonymous and non-anonymous reviewers commented on the initial proposal or versions of the manuscript as it navigated its circuitous journey to publication; *Beyond Recycling* owes its title to one of you. Routledge provided the opportunity to revive a project which seemed to have prematurely reached its end-of-life; I particularly thank Rebecca Brennan and Rosie Anderson. Maria Fernandez Marinovic produced all the original illustrations including the front cover, and collaborated with me on consideration of images throughout the final version of the book. Anne Chick provided my original entry point into recycling as a topic for research and subsequent critique; this book really began then. Others who contributed directly to the development and content of *Beyond Recycling*, through conversation or suggesting things I didn't know about: Gary Anderson, Anne Marie Fisker, Rosie Hornbuckle, Mike King, Dan Lockton, Christoph Lueder, Frank Millward, Chris O'Brien, Christopher Pett. The index was compiled by Non Lowri Evans.

INTRODUCTION

When did you last recycle? Perhaps it was today, or yesterday. What did you discard? Perhaps a plastic water bottle or piece of cardboard packaging. Do you know what happens to it now? If you took care choosing the right collection bin, you probably hope equal care is taken as your donated item begins its journey through the waste stream. Do you know where that journey ends? Is this something you think about when you recycle, or is it a case of 'out of sight, out of mind'?

Recycling has long been a pillar of environmentalism. To be considered an environmentalist, you have to be seen to recycle and encourage others to do so. An increase in collective awareness of the environmental consequences of our contemporary lifestyles has led to recycling going mainstream. Recycling has become a defining idea of our contemporary culture, and one of our most visible collective social practices. The recycling symbol of a triangle formed by three chasing arrows is inescapable. This heightened popularity of recycling is due to the fact that it presents an easy answer to the problem of waste. As we become satiated, and increasingly sickened, by the excesses of consumerism, we look for ways to address our unease with the visible costs of our materialism. We are troubled by the undeniable wastefulness of our lifestyles, in which everything is designed for our convenience and comfort, and largely disposable as a result. The plethora of physical stuff with which we fill and fuel our lives litters our immediate and global environment. We have come to realise that plastic never really goes away, but just moves on to other places.

Plastic packaging waste breaks up and degrades, but not in a way that the Earth's systems can reabsorb. Plastic fragments now fill the oceans; out of sight but no longer out of mind. And so we turn to recycling.

Recycling is presented uncritically as the right thing to do. In the workplace, at home and in public spaces we are told to recycle. Organisations, governments and cities have recycling targets, and we must play our part to ensure these are met. To not recycle is therefore to fail, and be judged. We are compelled to participate in recycling by a combination of moral responsibility, and the twin threats of financial penalty and social disapproval if we do not comply. Yet the mantra of the 3Rs – reduce, reuse, recycle – tells us that recycling should be a last resort, rather than a first choice, when it comes to dealing with waste. In our rush to recycle, we have forgotten that prevention is better than cure, especially when the proposed cure does not work. Recycling does not, and cannot, work as a meaningful response to the sheer scale of our production of waste. The volume of material we produce that requires recycling is not matched by the recycled products that we buy. There are few examples of successfully closing the loop on material use. Most recycling is 'downcycling' – turning waste into material of lower quality and value. It is not just the volume, but the rate at which we produce waste that is the problem, and recycling does not offer to slow this down. Recycled things can be deceiving, giving us false hope that they embody a sufficient response to what we throw out. Endless variations on the recycling symbol offer reassurance that all will magically be taken care of further down the line. We therefore feel good about recycling.

Recycling is a seductive idea, but flawed as a practical solution to the challenges of our material culture. Recycling – turning something considered to be waste into something with renewed value – is therefore ripe for closer examination. This book critiques the fact that we currently privilege recycling over approaches that could have much greater impact on the problem of waste. It argues that we must go beyond recycling if we are to seriously tackle the material costs of how we live.

In so doing, the book challenges the protected role of recycling in any collective transition to a state of greater sustainability.

A myth is a created story which helps us understand the world and our place within it. *Beyond Recycling* exposes recycling as an unsubstantiated contemporary myth, an element of the greater myth of sustainable development. It is argued below that sustainable development, rather than challenging the assumptions of our industrialised society and offering a route to a preferred, more sustainable future, simply reproduces the errors of the era of modernity. There is nothing truly sustainable about 'sustainable development', and the three pillars of economic, social, and environmental value do not provide a new framework for human flourishing. Recycling can be seen as an exemplar of ecomodernism, the common view that we will develop successful technological solutions to political and environmental challenges such as Global Heating. Frameworks such as the Circular Economy, which challenge the dominant linear approach of extracting, consuming and discarding materials as if their supply were infinite, are essentially ecomodernist unless they challenge the unsustainable structures of our society. This book therefore exposes the unsustainability of recycling.

Beyond Recycling explores unasked questions around recycling and its prominent position in contemporary thinking about sustainability. It examines why we have so wholeheartedly committed to recycling as a cultural project. Recycling has become a commonplace notion and widespread practice. The book considers recycling as a contemporary cultural idea related to, but not wholly defined by, our response to material waste. It seeks to reclaim recycling from the environmentalists and waste management specialists, to explore how the term is used in other contexts unconnected to material waste. *Beyond Recycling* offers a wide-ranging account of recycling as a contemporary ideology. It focuses on the period from the early-mid 20th century, when the language of recycling emerged, to the present. Earlier examples are discussed insofar as they help us to understand the present – there is, after all, nothing very new about much that we now call recycling. The book considers the contemporary politics of recycling, viewing it is a political practice at every level: individual, local, regional, national, global. *Beyond Recycling* is by necessity also about reuse, because to understand recycling we must understand its relation to reuse. The hierarchy of the 3Rs makes a distinction between the two, but they can sometimes be difficult to separate.

From being a proposed means to an end – what to do with our waste – recycling has become an end in itself, pursued for its own sake. Our collective turn to recycling has led us to believe that almost anything can and should be recycled, and remade as good as new. Recycling promises an alchemical restoration of material value, redeeming our waste and excusing our profligacy. It promises us continued mastery of the material world and its finite resources, without too much change to our present systems and practices. Yet as we collectively wake up to the Climate Emergency and a global crisis of unsustainability, and reconsider our relationship with the physical stuff which currently enables our lives, we must see that recycling is not enough. Recycling everything, even if it were possible, is not the answer. We must go beyond recycling if we are to take sustainability seriously. To do so, we must examine recycling thoroughly, to identify what it can and cannot offer us as we seek a transition to a liveable future.

Beyond Recycling shares its title and intent with an article which is equally sceptical of recycling as a potential solution to 'the waste problem' (Barr *et al.* 2013). The naming of this book preceded discovery of that article, but it is worth acknowledging the kindred perspectives. The earlier authors critique an exclusively positive environmental framing of recycling, advocating instead a more critical and social approach to the consideration of recycling, its practice and its merits. They also challenge the emphasis on individual action that is common to most discussions of recycling, arguing for a systemic and structural consideration. Whilst the earlier paper is located in the subject area of geography, this book comes from the field of design. Design provides the lens through which recycling is viewed in the chapters that follow. Products, communications, places and systems are all examined as they relate to the topic. Recycling was the author's entry point to engaging with design for sustainability, specifically a study of designers' attitudes towards, and use of, recycled materials. This book is the culmination of almost 20 years spent focusing on the role of design in pursuit of the goal of greater systemic sustainability, and the problematic role of recycling in that quest. It also shares its title with another influential earlier work, which advocated for longer product lifespans as a better option than product recycling (Cooper 1994). Then, as now, recycling was not the answer.

REFERENCES

Barr, S., *et al.*, 2013. Beyond recycling: An integrated approach for understanding municipal waste management. *Applied Geography*, 39, 67–77. https://doi.org/10.1016/j.apgeog.2012.11.006.

Cooper, T., 1994. *Beyond recycling: The longer life option*. London: New Economics Foundation.

1

LANGUAGE

A SHORT HISTORY OF RECYCLING

Transformation of materials in response to changing needs is not new. The Romans melted down broken glassware to make new products – a sensible strategy for making the most of a valuable and scarce material. They also refashioned sculpted busts of fallen emperors to resemble their successors, albeit with mixed success. 'Spolia' is the term for reuse of earlier building materials or decorative sculptures for new monuments. This was a widespread practice in late antiquity and the medieval West, as in the present cathedral of Cordoba in Spain, which previously served as a mosque. Easier to repurpose and adapt than start again. Spanish conquistadors melted down the ornate treasures of the Inca and Maya in Central America for more mundane use as gold coins. Benin bronzes, actually made from brass, were often created from metal brought to Africa by Europeans, melted down from the original form. English monarchs renewed and refashioned royal furniture as a matter of course, due to its embodied material value and the unsupportable cost of simply replacing it. During the Second World War, the railings of Londoners were requisitioned to be melted down for use in munitions, as part of a national salvage campaign, although whether this collected metal was actually used in this way is disputed.

The thrifty reuse of valuable material resources has long been practised by the grandest and most powerful, as well as the neediest. Whilst we might think recycling is a relatively recent invention, the reprocessing, reshaping and reusing of materials have occurred throughout human history. Some of the historical practices just listed might, however, be considered as examples of material *reuse*, rather than recycling. Melting down precious metals to make new decorative pieces does feel like recycling – the material is returned to a raw form and then remade into something new. But reworking furniture to fit a new fashion feels less transformative, and perhaps more like adaptation or reuse of an existing form. The border between recycling and reuse is something to consider.

The language of recycling is around a hundred years old. The first recorded usages of the terms 'recycle' and 'recycling' were in 1925 in a series of US patent applications relating to transformation of ammonia gas into other chemical compounds.

> With such cyclic operation, it is readily possible to recycle sufficient ammonia to insure that substantially all of the phenolic compounds will be driven off so that they may be recovered. (OED, 2009b)

> Even without such recycling of ammonia ... the process of the present invention permits the recovery of important amounts of phenol from the gas liquor. (OED, 2009c)

The term 'recycling' was thereby first introduced as a process of isolating and capturing a by-product – phenol – from a repeating chemical process. This was an industrial activity, requiring patentable technology. These first instances of use of the language of recycling establish it as a technical method of material reprocessing and extraction. Recycling became a national priority during the Second World War, when households were encouraged to save paper, cardboard, metals, rubber and any other materials that were becoming scarce. Propaganda posters promoted the value of salvage, and an ethos of 'make do and mend' to the British war effort. This is the basis on which recycling can be considered specifically as a 20th-century practice.

The language of recycling is relatively new, historically speaking. It does not simply give us new names for practices familiar from the past, but describes something new. The language of recycling includes other terms: 'recyclable' (capable of being recycled) was first recorded in 1963 (OED, 2009a); 'upcycle' and 'downcycle' (to recycle to a higher or lower material value, respectively) emerged in the mid-1990s (OED, 2014, 2018);

recyclate (used to describe both recycled material and material ready for recycling) has yet to enter the dictionary. The emerging language of recycling is rich. This range of new terms indicates we are making finer distinctions within our practices of material reuse, and that these practices are becoming more varied. The sophistication of our recycling exceeds that of our ancestors. What distinguishes Roman production and reproduction of glass from modern industrial polymer reprocessing is the scale, automation and complexity of the processes used. Roman re-smelting practices would look nothing like a modern recycling plant. Polymer upcycling is a chemical process involving finer molecular intervention than melting down glass for re-blowing. In scale and complexity, our recycling practices differ significantly from those of the past.

RECYCLING AND REUSE

Recycling of materials contrasts with direct repurposing and reuse of products. A gardener may reuse a transparent plastic bottle (a product) directly as a bell cloche to protect outdoor plants from cold temperatures and insect pests. The material of those same plastic bottles may alternatively be reused, with a much greater degree of adaptation, to make the filling for a fleece jacket, as a substitute for other natural or synthetic materials. The transformation of the plastic bottle is quite different in these two cases. To make the bell cloche we simply cut off the base of the bottle. To make the fleece filling we shred bottles in bulk and reprocess them into a new material for an entirely new purpose. The first is a domestic practice, available to anyone willing to be creative and make the most of a common everyday resource. The second is an industrial practice, possible only with specialist technical expertise and expensive equipment. One requires some degree of domestic organisation – storing empty bottles in the garden shed preparatory to their use. The other requires a concerted material collection programme with a guaranteed ongoing waste stream of sufficient quality and quantity. The material in the first case is simply 'plastic', and our interest lies in its immediate tangible properties of transparency and resilience. The material in the second case is polyethylene terephthalate (PET), chemical formula $(C10H8O4)n$, melting point 250 °C, density 1.38 g/cm^3. In the first case, the original product form of a transparent bottle is integral to the function of its second life protecting young plants from harm. In the second case, the original product form is irrelevant, and the material's chemical composition is of value.

The distinction between recycling and reuse is therefore based on the nature of the manipulation performed on the original product. Turning a two-litre plastic bottle into a cloche requires some input of energy in the form of human effort, and some transformation of its form by cutting off the base, but the composition and qualities of the material remain unaltered. This is clearly different to melting down a batch of bottles and re-forming them into a new material for an entirely new purpose, such as the filling of a fleece jacket. The cloche retains the form of the original product, albeit with some alteration. The fleece does not. Here we therefore draw a line between reuse and recycling in terms of some retention of *form*. There are degrees of this, however. Making a cloche by removing the base of the bottle is not as direct a form of reuse of a bottle as simply filling it with water and adding cut flowers, to transform it functionally into a vase. That simple act seems a pure example of reuse, leaving the bottle untouched and simply giving it new purpose. But what if we go further and remove the bottle's branded paper label? What if we discard the bottle cap, and perhaps even snip off the matching perforated ring of plastic around the bottle's neck? These transformations are not worthy of being called recycling, however, as they still leave the bottle recognisable, albeit with modifications. Recycling is a difference in the type, not simply the degree, of transformation. It seems to rest on the question, when is a plastic bottle still a plastic bottle?

A further distinction between recycling and reuse lies in ideas of a product's *newness*. In reuse the object itself remains largely the same. It has renewed function, but is not made anew. Inventively repurposing a product in a way unintended

Figure 1.1 Transformations of a plastic bottle.

by its designer – doing something new and different with it – gives that product a new function, but it remains the same product. Its *affordances* – the embedded codes and cues that tell us how to interact with and use an object – are simply reinterpreted. Reprocessing the material of the product in order to remake it in a different form is a more significant category of newness. We can think of products as temporary configurations of materials having some current benefit, or responding to some current need. This benefit or need may not last, however, in which case we may then consider the product merely the sum of its constituent materials. If the materials that make up a product are reprocessed and remade into a new product form, we feel more comfortable saying 'this is a new product' than when we simply employ an object for something other than its original purpose. We feel that the materials have a new identity in their new form, even if we know that they have been recycled from something else and so had a previous life. Reuse can be a simple domestic practice, such as the direct reuse of packaging as a storage container. We don't need to pay anyone to turn a plastic bottle into a flower vase for us; we can quite happily do it ourselves. We are likely to have to pay for a recycled object, however, as this will usually require specialist technical capability and some financial cost. Recycled products are therefore generally produced for sale.

There are thus several dimensions to our distinction between reuse and recycling: (1) Reuse involves the repurposing of an original object, rather than a transformation of its materials; recycling involves the reprocessing of the original product's materials to make something that we consider to be new. (2) Reuse is performed by users and consumers, perhaps on a one-off basis; recycling is an industrial activity done repeatedly and at scale. (3) Reuse of products is done for direct personal benefit; recycling of materials is a commercial activity. A middle ground between reuse and recycling, between handcrafts and material reprocessing, lies in experimental attempts to bring industrial recycling processes into the home, workshop or garden shed. Crude versions of volume recycling processes can be performed in the home – melting yoghurt pots and re-forming them as simple bowls – but they may be difficult to perform safely or effectively.

	REUSE	RECYCLING
DEGREE OF TRANSFORMATION	REPURPOSING OF AN ORIGINAL OBJECT	REPROCESSING OF AN OBJECT'S MATERIALS TO MAKE SOMETHING NEW
SCALE	PERFORMED BY USERS AND CONSUMERS, SMALL SCALE	INDUSTRIAL ACTIVITY, LARGE SCALE
AIM	FOR DIRECT PERSONAL USE	AS A COMMERCIAL ACTIVITY

Figure 1.2 Reuse versus recycling.

UPCYCLING AND DOWNCYCLING

We should be sceptical of the value of recycling plastic bottles into fleece jackets. Reprocessing polyester from plastic drinks bottles to make a fleece lining might seem a good example of recycling a waste material into a higher-value product. A fleece jacket is clearly a more sophisticated product than a water bottle, and also has much greater value and cost as a consumer product. Yet the material is actually being *downcycled* in this instance. The highest food-grade polyester, used for the bottle, is likely to be contaminated by the recycling process, making it chemically inferior in its second life in the jacket. The polyester in a fleece jacket is therefore of lower quality than that in a water bottle. The value of the material is diminished, rather than enhanced, by the recycling process. *Upcycling*, in contrast, is any process by which materials are purified and improved in quality, prior to reuse in new product forms. Where industrial recycling often damages materials such as polymers or paper by shortening their fibres and thereby lessening their desirable qualities, upcycling purifies materials and improves their quality. This is a restorative form of material recycling. Take the example of a typical plastic water bottle, which contains residues from a catalytic reaction that produces a potential toxin. The material in these bottles can be purified as they return through an upcycling system that removes the substance of concern and improves the quality of the polyester. Additionally, whilst the process of recycling often shortens a material's fibres, it is now possible to repair those fibres so that the material can be continually reused at the same level of quality. Rather than being an extended process of successive degradation to the point of unusability, plastics recycling now has the potential to become a process of restoration to as good as, or perhaps even better than, new (Makower, 2013). This is the optimal mode of recycling, the opposite of conventional downcycling.

The term 'upcycling' is more widely used, however, to describe individual repurposing and creative reuse of unwanted objects. Upcyclers seek to preserve every element of the value of an original object. Material and product form are both considered to have value. Recycling typically destroys and then remakes both, often to a lower value (downcycling). Upcycling-as-reuse avoids destroying any part of an object. The act of reprocessing a material, and in so doing destroying its shape in order to remake it as something new, is considered inferior to the preservation of that original shape in a newly made product. Better to stitch together used multi-layer juice cartons to make a shopping bag than melt them down to feed into a new production process to make more juice cartons. The value of the original product can be seen in terms of both the properties of its constituent materials, and its constructed form. Upcycling-as-reuse seeks to preserve both through simple manipulation of the original product such as stitching together, rather than an industrial process involving the fundamental transformation of its materials and their reconfiguration into the same or quite different products.

This mode of upcycling is low-tech and accessible to anyone with basic tools and the willingness and aptitude to use them.

Figure 1.3 Upcycling-as-reuse: shopping bag made from juice cartons.

This kind of creative reuse has entered the cultural mainstream as an aspirational pursuit for the modern homemaker, aligned to newly fashionable notions of thriftiness and craft. Upcycling is now commonly used to describe attempts at domestic improvement through the regaining or acquisition of handcraft skills and their application to salvaged or neglected products.

This mode of upcycling can be heavily invested with ideology. The drive to create with waste is celebrated for its authenticity. It valorises remaking both for its ends (reducing the disposal of waste), and also its means (fulfilment for the maker). Online communities share their creations, to mutual acclaim. To the experienced eye, upcycled product archetypes constantly reappear, for example lightshades which are an array of plastic bottles, and ashtrays made from flattened melted beer bottles. Whilst the outcomes can be repetitive and derivative, the purposes and processes, both mental and physical, of upcycling are more interesting. It is the process of doing-it-for-yourself in response to your own needs and constraints that is of most value. Online community groups prioritise original upcycling projects over how-to-do-it guidance on emulating existing products. The presumption is that imitating someone else's response to their own needs will be less satisfying, and less worthwhile, than developing your own. Better to learn from the motivations and processes of someone else, rather than simply mimic their outcome.

Other examples of adding perceived value to an existing object through visual augmentation are the 'Re-Directed Art' paintings of Canadian artist David Irvine. Discarded prints of oil paintings are salvaged from thrift stores and elsewhere and embellished with familiar characters from popular culture. Conventional bucolic scenes now feature Darth Vader, Dumbo or Pac-Man, or surreal additions such as huge flying steaks or cycling birds. The revitalised prints are renamed to reflect the new subject matter, for example *Approaching the Finch line. Night on the Town* sees the Stay Puft Marshmallow Man from the film *Ghostbusters* looming at massive scale over a portrayal of a Parisian evening street scene by the French painter Antoine Blanchard. Irvine himself uses the term 'upcycling' in relation to his practice, reflecting common acceptance of the term to describe the re-presentation of existing raw material such as a painting which no longer has perceived value. He strives to either emulate the visual style of the original work, so that his embellishments appear as uncanny additions to the depicted scenes, or else he aims for jarring contrast with the original. In either case, Irvine is careful not to obscure the signature of the first artist, when there is one, whilst always adding his own. In this way the dual authorship of the renewed object is made clear, as the original artwork is still perfectly recognisable in the altered version. Irvine responds directly to what is already there in the paintings he selects, establishing a dialogue between himself and the first artist through his over-painted additions.

British artists Jake and Dinos Chapman work in a similar way, but with much higher stakes, in their series of 'rectified' original prints by the Spanish Romantic painter and print-maker Francisco Goya. In this case, the altered works are not obtained from charity shops or thrift stores, but represent a central part of the oeuvre of one of the most revered European Old Masters. The original prints, known collectively as *The Disasters of War*, depict scenes of gruesome violence and suffering in a series of conflicts in the early 19th century. The Chapmans defaced each image with a deliberate lack of reverence for the original work; 'we've gone very systematically through the entire 80 etchings, and changed all the visible victims' heads to clowns' heads and puppies' heads' (Jones, 2003). Whilst David Irvine's paintings are intended and received as playful reinventions of discarded original works, which had otherwise lost their value, the Chapmans are widely accused of vandalism. The fact that they are the legal owners of the artworks they deface is not seen as giving them the right to do so. The contrasting reactions to these two examples of alterations to purchased paintings and prints are interesting in what they say about our attitudes towards the remaking of artworks as cultural objects. The acceptance of Irvine's work, apparent in universally positive press coverage, suggests that we are quite happy to see redundant cultural objects, in this case discarded prints of unfashionable landscape paintings, refreshed by the painter's addition of contemporary cultural references. The widespread condemnation of the Chapmans' similar alteration of artefacts from the Western canon suggests that there are clear limits to our willingness to accept such reinvention; 'Goya's *Disasters of War* is not some dry old relic no one cares about – it is a work that has never lost its power to shock' (Jones, 2003). We are therefore shocked a second time by the disrespect shown by the later artists. Their use of the term 'rectified' to describe their reworkings of Goya's prints also upsets us, with its implication that they are in some way correcting deficiencies in the originals. There is clear intention to provoke in the Chapmans' use of language in relation to what they have done – *Insult to Injury* is the title they give to the series of prints. They seek to harness the power and moral anger of Goya's original works and redirect them towards the forces of contemporary consumer culture, which they attack through their work. In turning the victims into clowns and puppies, the Chapmans add a layer of grotesque to Goya's original depictions, thereby ridiculing the sentimentality and commodification of much contemporary popular culture. Whilst equally humorous, albeit much more bleakly, their intent is more serious than that of Irvine. The hostile response to one but not the other of these artistic interventions suggests that we are uncomfortable in recycling objects of significant apparent worth.

NON-MATERIAL RECYCLING

As well as describing the ways in which we reuse physical materials, the language of recycling has metaphorical value in relation to other types of material. 'Recycling the weather' claims a sign at a nature reserve, explaining how rainwater is collected from the roof of the visitor centre and reused in the toilets. The rainwater is not materially transformed, but simply diverted for an alternative use. Rainwater might be seen as a product of weather, but it is not the weather itself. There is therefore no real recycling going on here. Use of the term does, however, suggest that we are comfortable with the idea of recycling a complex and conceptual system such as weather.

Coin and banknote recyclers are machines which automate the sorting of currency, enabling large volumes to be counted quickly and made ready for further use. In this case, the value of loose currency is recycled through a handling and sorting system which makes it newly available, without it undergoing any material transformation. A more significant recycling of money is done via illegal 'laundering', by which the origins of 'dirty' money obtained illegally are obscured by passing it through a complex sequence of banking transfers or commercial transactions. This cleaning process returns the money to the launderer, to allow its legitimate and unencumbered reuse. Through this process, money may change form multiple times between physical and digital, its value retained throughout in the same way that the value of a physical material can survive its transformation and embodiment in a succession of products. Just as the quality of a material may diminish through successive phases of recycling, for example as the fibres shorten due to the rigours of reprocessing, so the original value of a sum of money may lessen as the intermediate costs of laundering are subtracted.

In sport, teams of rugby players use rucks and mauls to recycle the ball after a tackle, a tactic used to retain possession and gain territory from the opposition. When a player carrying the ball is held up by an opponent, or goes to ground, this is an invitation for a pile-up of bodies, with the purpose of opening up the next phase of open ball-passing play. In a maul the ball is held off the floor, and all the players must try to stay on their feet. In a ruck, the ball is on the ground and must not be handled by any of the players – they use their feet to ruck the ball backwards. The term 'recycling' is used to describe the transferral of the ball from a dead to a live state, and the continuation of the game via a coming together of heaving bodies.

Possession of the ball can also be recycled in football. Recycling possession involves an attacking team moving the ball into a deeper area to a player who is providing cover, so that she can move the ball to restart build-up play leading to a new attack on the opposition's goal. By recycling possession, an attacking team makes a tactical retreat when there are

Figure 1.4 Recycling the ball in rugby.

insufficient opportunities to penetrate the opposition's defence. Recycling the ball was a feature of the patient attacking play associated with Barcelona FC, the dominant European team of the first decade of the 2000s. In this style of play, possession of the ball and encampment in the opposition's half of the field are maintained to an exceptionally high degree. The opposing team is forced into a permanent extremely defensive position, creating huge pressure on their penalty area. The ball is played around and in front of a massed defence until a clear opportunity for a more incisive pass into the penalty area, and a more confident attempt at goal, arises for the attacking team. It takes a supremely confident and skilled team to turn away from an opponent and look to attack in a new and more productive area.

In these sporting examples the ball remains materially the same as it enters and leaves the recycling process, yet in each case it is symbolically transformed and a new phase of play is entered into. In this sense the game underway is renewed, even if nothing is physically altered, and all is materially as it was at the start of the game.

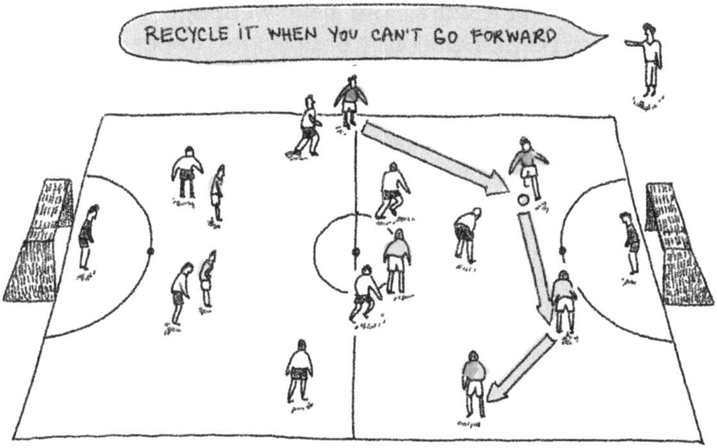

Figure 1.5 Recycling possession in football.

Information and communications technology handles immaterial content – the text and images that fill the virtual spaces which we access via our digital devices. This content lacks any physical presence and weight, yet has cultural and economic value which makes it worthy of recycling or reuse. Copying someone else's content without giving due credit is the cardinal sin of the digital age. The plagiarist, whether accidental or deliberate (often, no distinction is made), is as guilty as when she rides off on a bicycle that isn't her own. This is despite the fact that technology makes immediate reappropriation and transmission of digital content so easy, dissolving conventional notions of material ownership. The progressive approach to intellectual property rights contained in the Creative Commons licensing system adapts traditional notions of the ownership of content, by encouraging the free dissemination and adaptation of a participating author's work by others – an example of recycling of immaterial content.

Another claim for the recycling of virtual content is made by a computer wallpaper image that declares 'made from 100% recycled pixels'. The digital reconfiguration of a screen display which takes place as a new webpage loads on our computer does not represent any material transformation of the computer's hardware. The new webpage is constituted on the screen – instantaneously or in stages, depending on our internet connection speed – as each of the many thousands of individual pixels switches to display its own new image. These individual images are then combined by the visual processing capacity of a human brain into a new composite image which is perceived as a new webpage, occupying the entire screen. There is no physical transformation – the pixels remain structurally unchanged. Yet the data which they present as images does change, combining to make the new composite image that we see as the new webpage. This feels like recycling to the viewer now looking at a new webpage, with its new content and meaning. This is akin, conceptually, to the disaggregation and recombination of the molecules of a polymer as a water bottle is reprocessed to form the fleece lining of a jacket. When we refresh a current webpage, rather than load a new one, things get even more interesting. In the case of a reloaded webpage, the same whole-screen image is remade identically, the same pixels carrying the same image as before, with the possible exception of the clock at the bottom of the screen, which updates each minute to give the current time. Not only is there no physical transformation of the screen in this case, but the data presented by each pixel is also identical to that before the page was refreshed. This feels more like reproduction than recycling – an identical remaking of the same image, a form of non-biological cloning.

The most obvious use of the language of recycling in computing technology is of course sat right there on our PC desktop screen: the Recycle Bin to which we consign our unwanted files by pressing the 'delete' key. This is in fact a holding pen, a special directory in which files sit in limbo prior to their final extermination when we select the command 'Empty Recycle Bin'. Files thus deleted can often be recovered from this oblivion by someone with the right technical skills, but for most of us an empty Recycle Bin icon denotes the loss of those files forever. Another way to use the Recycle Bin is to store documents such as letter templates that we routinely repurpose – where better to store these than a folder designated for recycling?

Figure 1.6 Computer desktop Recycle Bin.

Another specialist use of the term in computing is the recycling of application pools, whereby computer servers that run systems or websites have their memory periodically flushed, or recycled, to clear out stale content. A daily recycle ensures that the server doesn't run out of memory and continues to function effectively. This recycling routine may see the server briefly running more slowly, but is preferable to running out of memory, which would mean the system or website not working altogether. An 'out of memory' error message on a website means that the recycling strategy isn't quite right. Recycling is a maintenance issue, and denotes a regular process of upkeep to ensure acceptable continued performance. Recycled here means cleaned, a regular necessity to return a server to an optimal functioning state, albeit temporarily. This example stands out in the current discussion, as it relates to the repeated cycling of a configurational process rather than any material or immaterial transformation. Looking at the server itself we would little suspect that any recycling had taken place.

The contents of electronic files are also endlessly recycled, or even directly reused, by their authors. Writers adapt the content of previous articles for new readers, reformulating and re-presenting ideas in a new format for new consumption. We may do this transformational work on our own past writing, or that of others. Plagiarism arises when due credit is not given to original sources. Self-plagiarism is the act of reusing one's own existing published or submitted material without adequate acknowledgement or development. Yet all writers inevitably build on what they have written previously. Every new work emerges from or responds to what preceded it. Political speech writers recycle earlier scripts, even across party divides, ensuring that their audience eventually gets the message that the politician wants to put across. The researcher's unsuccessful application for project funding will be recycled for a future bid, which she hopes will be successful. In our era of accelerating networked media it is often in fact very difficult to identify the origin of an idea or expression to one particular source. In this sense all ideas are at least partly recycled, communally generated and jointly owned. The collective stories we tell ourselves to make meaning of our lives, which help constitute our culture, are constructs made from the content of preceding narratives. Urban myths, our modern folktales, express long-standing assumptions and desires, and are transmitted and transformed as an oral tradition. Whilst the recycling of ideas may be viewed negatively, as lack of originality and new thinking, the transformation of existing material into new forms, perhaps with only slight differences from what has gone before, is the norm for written and verbal texts. Successful writing resonates with what we are ready to read or hear. It does so by extending or expanding what is already familiar, giving the impression that if it is not entirely new, it is at least refreshed and more useful than what we had before.

A consideration of the idea of the cultural industries is helpful here. These are 'industries that make texts', where a text is any type of cultural work or product which is concerned primarily with the production of social meaning rather than functional value (Hesmondhalgh, 2013, p. 30). Films, recordings, books, comics, images, magazines, newspapers are all cultural texts. The hardware and software used to deliver these texts – television sets, radios, computers – have primarily functional value. Those working in the cultural industries to produce these texts have 'symbolic creativity', the outcome of which is both intangible and ambiguous – the meaning of a cultural text will always be contestable, depending on the perspective of the viewer. A worker in the cultural industries, as a producer of texts rather than things, is therefore engaged in the reallocation of meaning to existing tangible materials. Our consumption of cultural products is also essentially non-physical.

> Cultural commodities are rarely destroyed by use. They tend to act like what economists call 'public goods' – goods where the act of consumption by one individual does not reduce the possibility of consumption by others. If I listen to a CD, for example, that doesn't in any way alter your experience of it if I pass it on to you. (Hesmondhalgh, 2013, p. 30)

Nothing is physically altered or destroyed when we consume a cultural product – an artwork is not affected by our looking at it. Its meaning, however, may change, and repeatedly, as it is viewed by more and more people from diverse perspectives. We can think of this constant reinvention of the meaning of a cultural text as a mode of intangible recycling. If we acknowledge a role for the viewer in creating, or at least contributing to the creation of, the meaning of an artwork, the artwork is never finished – there will always be someone new to view it, and make new meaning of it.

Practitioners of symbolic creativity are likely to engage in the rewriting of cultural texts of all kinds. Artists commonly explore recurring or dominant ideas throughout their working lives, producing multiple variations on the same theme – Edvard Munch's scream, Van Gogh's sunflowers, Monet's water lilies. Designers are also often simply refreshing or updating existing or established products. In fact most design begins with a consideration of existing products, consciously or unconsciously, which are then reworked, more or less creatively. This can be seen as a form of recycling. Graphic designers are therefore constantly recycling existing visual identities and graphic devices, a practice which the book *Recycling and Redesigning Logos: A Designer's Guide to Refreshing and Rethinking Design* celebrates (Hodgson, 2010).

Sound recordings are another notable category of cultural text. *Recycle* is the title of a 'restoration' of the early singles releases of the Manchester bands Joy Division and New Order, for unofficial release as a virtual boxed set (an oxymoron which we have come to accept in the age of the streamed on-demand television series). This fan-based project makes available rare and out-of-print recordings, converting them from their original physical vinyl format to higher-quality digital files for download. New treatments of the recordings themselves and their accompanying artwork sought to improve on the originals:

> All tracks were taken from the best/earliest possible sources to avoid modern mastering techniques which crush the dynamics. Tracks sourced from vinyl have been carefully cleaned and EQ levels have been tweaked for consistency. The artwork was scanned at the highest possible resolution and the type was reset when possible using the original fonts. (Metzger, 2010)

A visual packaging concept for the collection uses a version of the famous Mobius loop recycling symbol (discussed in detail in Chapter 3 of this book) as its central graphic device. In striving to reconstruct versions which are better than the originals, the project should perhaps be retitled *Upcycle*.

Recycling is the title of a short story in which a newly single mother realises that she must move on from her misfortunes:

> She reminded herself that she had told him to leave. She had done nothing wrong, and now she had to start living again. Recycle herself. (Livingston, 2011)

The lead-up to this moment of personal resolve is her daughter's fixation on the new recycling collection scheme introduced by the local council, which she has also discussed with her teacher in school. The central metaphor of the story is, if waste materials can be recycled, why not wasted people too? The story makes an overt correlation between these two forms of redemption. The road to personal transformation and renewal – moving on from a neglectful lover and the associated self-destructive behaviour – begins with the appropriate disposal of a vodka bottle (which the mother presumably now no longer needs to get her through the day). This is personal upcycling, as the mother emerges empowered from a process of reflection and reconsideration. Identity is reprocessed into a higher-value version of the self, or at least a version of the self that values itself more highly.

A similar focus on the recovery of dormant or wasted human talent and capacity is at the heart of the social enterprise movement. In this context, human capital is seen as an asset alongside financial, natural and cultural capitals, yet one which is frequently squandered or unrecognised in the conventional economy. Joblessness is a primary cause of homelessness, crime, poor environment, obesity, mental health issues, including depression. Social enterprises often see meaningful employment as a valuable end in itself, in redeeming wasted human potential for achieving personal fulfilment and making a wider societal contribution. Tree Shepherd is a London-based project which uses a metaphor of cultivation to describe its mission to nurture enterprise and employment opportunities in neglected parts of society, transforming job-seekers into job-makers. It identifies lack of work and its consequent hopelessness as a cause of serious social and health issues that can accumulate and become endemic. The founder of Tree Shepherd, Colin Crooks, was previously founder of a social enterprise which sought to address this same agenda via a business focused on the collection and redistribution of unwanted office furniture. GreenWorks sought to employ people who found it difficult to enter the labour market due to a lack of qualifications or having criminal convictions. The recycling of waste furniture and office equipment served the broader agenda of repurposing dormant human potential, giving training opportunities and employment to those whom society saw as having little to offer. GreenWorks was a recycling business in more ways than one, applying the idea as a positive force for more effective use of physical and human resources (Crooks, 2012).

The same metaphor of personal recycling is applied to the difficult transition to retirement from work, with its concerns and anxieties over what to do with the years to come. The negative associations of retirement are replaced by the altogether more positive associations of recycling as a form of personal renewal. Fears about retaining our usefulness when our working life ends can be allayed by 'self-recycling'. The cover of the book *Don't Retire … Recycle Yourself* (Kaloides, 2006) once more uses the classic chasing-arrows recycling graphic device to advocate self-reinvention as an alternative to an unplanned and empty retirement. The notion of recycling is used here to express the purpose of making a new start, and finding new goals.

Whilst we may be metaphorically recycled in life, we are inevitably physically recycled in death. Cremation has replaced burial as the most common way of treating the bodies of the dead in Western societies. We can see this as dropping down the waste hierarchy, to the incineration option which sits below the familiar 3Rs of reduce, reuse, recycle. Historical practices of natural burial which are akin to recycling are, however, being rediscovered by the ecologically minded. Natural burial places the body in the soil in a manner that encourages decomposition, allowing it to recycle naturally. Conventional burial seeks to slow down bodily decomposition via embalming and use of a hard coffin to shield the body from the processes of earthly reprocessing. In natural burial a body is wrapped in natural fibres and placed directly in the earth to biodegrade. The re-emergence of natural burial is perhaps representative of a more general collective turn to traditional practices around

Figure 1.7 Sky burial.

waste and its disposal, motivated by environmental awareness and a desire to live, and die, with a lower environmental impact. The contemporary recycling agenda, as we have seen above, embraces a range of activities which would be recognised by previous generations as sensible management of limited resources. If our grandparents were motivated by thrift, we are more likely to be driven by environmental ideology.

The physical recycling of a body in death can take place through the slow biological processes of natural decomposition, or be assisted in traditional sky burials by direct consumption by birds and animals. Cremation is a more dramatic process of downcycling, in which the body is simply burnt and its ashes kept for some form of observance by loved ones. These physical recycling processes are to some extent slowed by conventional funerary practices such as embalming. They may even be halted by cryopreservation, low-temperature preservation of the body until such time as healing and resuscitation may be possible. We do not yet have the means to revive or restore life to a dead person, at least outside the pages of Mary Shelley's *Frankenstein*, in which a nameless monster is created from an assembly of non-matching recovered human body parts, reanimated by electricity. Common to all religious belief systems, however, is some faith in the continuation of a personal spirit or soul after physical death. The monotheistic Western Abrahamic religions – Judaism, Christianity, Islam – maintain that the soul continues in a heavenly afterlife. The body is abandoned as the soul progresses to a higher non-corporeal state. The polytheistic Eastern religions of India, China, Japan and South-east Asia, in contrast, commonly maintain that the soul is reincarnated on Earth following bodily death, beginning a new life in a new body. The old body is abandoned and a new one inhabited, as the soul begins a new cycle of earthly life. This continuity of life and death is characterised in these faith systems as a continual process of *becoming*, as the soul proceeds through corporeal states of higher or lower status, depending on the moral quality of the previous life's actions. This is the core idea of process philosophy, that all substance is always in a state of becoming and therefore part of a process of continual recycling and re-formation (Rescher, 2000). The classic philosophical position on substance, fixity and change comes from Aristotle: if I fall sick I remain the same person, and my sickness is only a temporary change. Process philosophy instead proposes that constant change is the permanent state. This has implications regarding notions of recycling and value.

The view that everything is always in a state of becoming, and never becomes final or complete, runs counter to a capitalist consumer model of commodification, sale and ownership. The primary carrier of economic value in this system is the product which is bought and sold. Yet a product is only a temporary combination of elements and materials in a form which has some immediate benefit, or which fulfils some current need. If that benefit is no longer given, or that need no longer met, then the product becomes valueless and redundant, it becomes waste. Recycling is the waste management approach which seeks to recover the inherent value of the component materials that go to make up a product, for reconfiguration in a new form which has ongoing use and value, thus saving it from disposal in landfill. Recycling asks us to view products as temporary milestones in the evolution of the more fundamental materials from which they are made. Recycling, as an idea and a practice, therefore, embodies the core concept of process philosophy that all things are always in a state of flux and transformation. The fixity of the product is temporary, a momentary freezing in time of an eternal process of material transformation. A pristine new product becomes tarnished and worn as soon as it is used. If all production and consumption is synthesis and recombination, conceptually and physically, then all is recycling. Nothing is ever entirely newly made if we look closely enough.

RECYCLING MORALS

Recycling has become a dominant social norm, expressed via the mantra of the 3Rs. There are many recordings available online of the song 'Reduce, Reuse, Recycle' being performed by the musician Jack Johnson with different groups of schoolchildren (Explore Documentary Films, 2009). Johnson is well known for encouraging pro-environmental behaviour amongst his audience. Whilst he would not claim to have created the 'reduce, reuse, recycle' refrain, the rest of the song's recycling-specific lyrics are Johnson's own. The song's musical form is, however, directly appropriated from an earlier children's song, 'Three Is a Magic Number' (Dorough, 1973). It is fitting that a song promoting recycling is itself recycled, and makes use of a widely known refrain of unknown origin – no one really knows who first said reduce, reuse, recycle.

Promotional efforts encouraging us to recycle usually emphasise that it is obligatory, but also reassure us that it is easy. Recycling is encouraged by carrot and stick. We must do it, but we won't find it too difficult. Indoctrination into a view of recycling as a civic duty begins early. There are numerous books and gaming apps for the youngest children which promote recycling as a desired social norm. In *Recycling Fun!*, the famous Peppa Pig and her family dutifully collect and sort all their recyclable household waste for recycling. This is a shared endeavour in which every family member plays a part. The environmental credentials of the recycling mission are, however, undermined by the fact that the family travel to the recycling centre by car, in what appears to be a special journey made just for that purpose. If so, this exemplifies a commonplace naivety as to the real, rather than perceived, value of our recycling efforts. The fuel and energy consumed by the car journey would exceed any potential resource savings from the diversion of materials from the waste stream. This is unthinking recycling, focused only on materials recovery, with no consideration of any other impacts which might arise from how it is done. The motivation to recycle the family's bottles, tin cans and newspapers comes not from any discussion of the environmental impacts of their manufacture and disposal, but rather the fact that the bin collector makes a lot of noise when he collects the household rubbish early in the morning. The motivation is in this case wholly social. As pro-recycling propaganda, this Peppa Pig story demonstrates a lack of holistic thinking common to much of our waste recycling. Any recycling effort is

Figure 1.8 Recycling as an end which justifies the means.

generally presented as an inherently good thing, without any deeper consideration of whether it is actually worthwhile, at least in terms of environmental sustainability. Disturbance from noisy bin collectors is an unusual motivation to recycle. It is also a poor justification for recycling.

Similar simplification is evident in recycling-themed gaming apps aimed at young players. These add rubbish collection and sorting activities to familiar gaming formats, using familiar game play conventions to make the separation of different types of waste entertaining. The degree of challenge of course differs according to the recommended age of the player. Those aimed at the very young (four years and over) may only require objects to be diverted into waste receptacles, or sorted by colour; 'Built-in help will instruct beginners in the noble art of trash sorting' (Anderson, 2012). Those aimed at slightly older players typically require differentiation between types of waste. The game *Trash & Fun* uses an attempt-and-error format to allow players to learn how to distinguish between the most common categories of municipal waste. Aimed at six- to eight-year-olds, the game builds up degrees of difficulty by introducing additional categories of waste at successive levels: Level 1 = paper and glass; Level 2 adds metal; Level 3 adds organic; Level 4 adds plastic; Level 5 adds wood. With the addition of each new category the difficulty increases of successfully diverting items of waste falling from the top of the screen into the appropriate bin, before they hit the ground. *Trash & Fun* has clear educational intent, emphasising the need to 'build a cleaner world' by learning to differentiate kinds of trash and the principles of separated waste collection (Trash & Fun 2014). The game uses the conventional colours of waste sorting – red for plastic, blue for paper, and so on. The hope is that successful players of the game will take their learning into the real world, and recycle their actual material waste in the same way. A variant on the same gaming format is aimed at university students, who presumably did not have the benefit of playing such games when they were younger (Manchester Metropolitan University, 2014).

Figure 1.9 Recycling gaming app.

The game *My Green School*, again aimed at players aged six to eight years, presents a more realistic account of waste collection by including landfill as an additional option. It also elaborates on the recycling sorting game format by locating recycling within the framework of the 3Rs – reduce, reuse, then recycle. Waste sorting is itself located within a broader mission of encouraging children to value and seek to preserve their natural environment. Recycling is thus one of a much wider set of actions which the player of the game is encouraged to take, such as creating a garden and reducing consumption (Konky, 2015). *My Green School* is apparently based on a real experiment in a real school, and resonates with the recent increase in sustainability-themed education in schools, certainly in Britain.

Another gaming format to be given a recycling makeover is the truck-driving simulator. Familiar game play is embellished by incorporating the kerbside collection of refuse. Successful return of a full truck to the depot is rewarded by progression to the next level of the game, and an upgraded truck with which to negotiate additional obstacles and driving hazards. These games simply involve driving a truck which happens to be engaged in recycling collection. They do not improve our knowledge of how or why the recycling of different materials takes place. As such they add to players' awareness, but not understanding, of recycling.

The examples just given illustrate that the language of recycling is one of duty, even morality. Recycling is a civic responsibility that we feel obliged, and are increasingly compelled, to fulfil. The theme of a national Recycle Week in the UK was 'Recycling at Home and Away', reminding us that this is not something we should be doing only at home or in the workplace, but everywhere we go. Recycling should be a state of mind. In some cases, recycling seems to be a religion, judging by the fervour of its most committed advocates. Placing waste items out on the street for recycling collection in colour-coded bags and crates is a corrupted modern form of quasi-religious observance, with the same associated risks of hypocrisy and sanctimony. Empty wine bottles become votive offerings, displayed outside our homes for our neighbours to see – we perform recycling for a public and municipal audience. Recycling can become revelatory and competitive, as we judge others and are judged in turn by what we place on the pavement for absolution.

Our participation in municipal recycling programmes brings us no direct personal benefit, beyond feelings that it is the right thing to do and of a duty fulfilled. Whilst there may be no tangible reward for recycling, there may be very real penalties for not doing so in the form of disapproval of friends and neighbours and financial penalty from elected authorities. Like many of the things we do voluntarily or for no apparent personal gain, recycling has ethical, even moral, overtones. This contrasts with the much more tangible basis for traditional recycling, which is economic. Scrap dealers who crush abandoned cars and resell the metal are primarily helping themselves rather than being motivated by ideology or social norms. Only recently has recycling become an overtly moral agenda. The sustainability team at a university unashamedly uses this language when it identifies the legal and moral obligation to correctly dispose of all waste generated by the university. A tote bag bearing the slogan 'Recycle or Die' in bold block capitals, purchased from a mainstream high street store, is further evidence of this moral imperative. Whether or not we see any irony, either deliberate or unwitting, in this threat displayed on a flimsy shopping bag will depend on who is carrying the bag and in what context. The slogan is in any case a cultural marker because it identifies that to not recycle is in some way to transgress. This oppressive aspect of our contemporary commitment to recycling is referenced in a song lyric by O'Hooley & Tidow (2010), which suggests we ourselves are reduced, reused and recycled, our spirits suppressed by an overbearing and enforced morality. Whilst the mantra of the 3Rs is unproblematic for Jack Johnson, and serves as a catchy chorus, here it is less benign, and represents compliance to constraining social norms. The 3Rs do not deal just with our material waste; they also deal with us, as their observance becomes a social and moral obligation. Johnson presents this as an empowering opportunity for personal advocacy, even activism. O'Hooley &

Tidow detect the opposite – disempowerment and a loss of agency – in the pressure to adhere to an enforced orthodoxy. We ourselves are reduced-reused-recycled, and become like the stuff taken away by the council. Compelled to sort our various empty bottles and tins into specified categories, to then be dealt with on our behalf, we feel unwelcome action upon us by the social and political system. Such miserablism is far from the optimistic environmental advocacy of Jack Johnson.

There is also a question of morality in relation to what we should or should not seek to recycle. Various approaches to the treatment of human bodies in death are discussed above in relation to contemporary practices of waste management, including recycling. Frankenstein's fictional creature represents a singular attempt to recover the value from dead human bodies now viewed as material waste. A proposal to do so on an industrial scale is given in the artist project *SlaveCity*, which proposes a system to recycle humans. This self-styled 'ecological design' presents a production system that emulates nature's in-built recycling model:

> Old, crippled, sick and bad tasting people will be recycled in the biogas digester. Healthy, not so clever people will be recycled in the meat processing factory. Young and very healthy people will be able to take part in the organ transplant program. (Atelier Van Lieshout, 2009)

This fictional proposal is directly inspired by extermination camps in Second World War Europe. In these hellish places, the living bodies of undesired people were considered an inconvenience which necessitated the construction of industrial facilities for killing and subsequent disposal. Whilst some resources were recovered from those bodies – spectacles, rings, hair – the bodies themselves were considered valueless and simply discarded. *SlaveCity* takes this principle further by treating human bodies as resources to be harvested and fully exploited without any regard for conventional notions of morality or ethics. What jars most is the project's direct use of the language of sustainable development:

> *SlaveCity* is the first 'zero energy' town of its size in the world and functions without imported mineral fuels or electricity. The energy needs of the

> city are covered by using biogas, solar power, wind energy and bio-diesel. Everything is majestically recycled, even the participants themselves, whose vital organs are destined for transplantation instead of decaying into dust. Since no waste products are produced, *SlaveCity* is a green town that does not squander the world's limited resources. (Vanstiphout, 2007, p. 39)

SlaveCity takes the principle of resource efficiency to its extreme, proposing an organisational system in which humans have value only whilst they are able to be productive. Once they stop being so, they are to be physically recycled for useful constituent material: blood for transfusions, organs for transplant, meat for food. Other works by the same artist studio repeat this theme. Works titled *Recycling* and *Vomiting Trio* depict the direct use of human bodily wastes as sources of human nourishment, improving on the efficiency of using those wastes only as fertiliser for the production of crops for subsequent use as food. The work of Atelier Van Lieshout functions as speculative or critical design, proposing an alternative version of the present which forces us to reflect on the validity and appropriateness of the status quo. The forces of rationalisation and efficiency that are the dominant imperative of capitalism are presented in extreme forms which are nevertheless not unimaginable versions of the present. Atelier Van Lieshout believes it would take only ten years (the project-as-proposal began in 2005) to obtain the necessary permits and to influence political opinion sufficiently to allow the first element of *SlaveCity* to be built, such was its potential financial profitability and return on investment (Allen *et al.*, 2011, p. 300). This only adds to the disquiet that the works create. Is this what recycling might look like if we continue to pursue it in unquestioned isolation from other considerations?

A less dystopian bodily recycling programme has been promoted via the slogan 'Recycle Your Organs'. In this case organ donation was presented as a recycling programme seeking to increase awareness and encourage people to become donors, invoking our sense of social responsibility (Recycle Your Organs, 2014). Responding to a lack of donated human organs from the recently deceased for use in the still living,

Figure 1.10 SlaveCity: Dentist, by Atelier Van Lieshout.

this campaign exploited the established cultural capital of the notion of recycling, and its famous visual symbol of the three chasing arrows. The gruesomeness of Atelier Van Lieshout's proposals is replaced by an appeal to civic responsibility. If we happily donate our domestic waste for recycling or reuse at the end of its useful life, why not do the same with our bodily organs at the end of our own lives? This seems a reasonable extension of the logic for recycling from one domain into another where the benefits are actually much greater. Recycling becomes increasingly worthwhile as the value of the donated materials increases. Whilst most domestic waste such as food packaging is of relatively low material quality and reuse value, in donating our organs we have the potential to save or prolong another human life. If we are to recycle anything, it should therefore be our organs as the most valuable parts of our own bodies. Increase in public support for organ donation appears to follow the introduction of an opt-out system, as a replacement for an opt-in system. Whilst both systems are voluntary, the organ donation consent rate in one UK nation rose from 58% to 77% over three years following the introduction of opt-out (NHS Blood and Transplant, 2019, p. v). This compares with a 67% consent rate amongst all UK nations combined (NHS Blood and Transplant, 2019, p. 2). Legislative change enabled more people to agree to donation, and more transplants to take place. As with recycling our domestic waste, we need encouragement to recycle ourselves.

CONCLUSION

In one sense there is nothing new about recycling. We have been transforming materials in response to changing needs since the beginning of human history. Philosophically, all substance is perpetually in a state of becoming, and therefore part of a process of continual recycling and re-formation. What is new is how we do recycling, aided by technological systems that enable action at industrial scales. Whether this is a good thing is examined in Chapter 4 Sustainability. The profile of recycling has also risen, and it has entered the ways in which we talk about many things that are unconnected with physical materials – we recycle ideas, texts, actions. Recycling is, however, still primarily an agenda for what we do with physical materials, embodied in the stuff with which we fill our lives. The next chapter explores these manifestations of recycling in the form of products and things.

REFERENCES

Allen, J., Betsky, A., Laermans, R. and Vanstiphout, W., 2011. *Atelier Van Lieshout*. Rotterdam: NAi Publishers.

Anderson, D., 2012. Trash it [online]. Available from: https://appadvice.com/app/trash-it/499841372. [Accessed 1 September 2020].

Atelier Van Lieshout, 2009. Cradle to Cradle [online]. Available from: https://www.ateliervanlieshout.com/works/. [Accessed 1 September 2020].

Crooks, C., 2012. *How to make a million jobs: A charter for social enterprise*. London: Tree Shepherd.
Dorough, B., 1973. Three is a magic number. In: *Schoolhouse Rock: Multiplication Rock*. Capitol.
Explore Documentary Films, 2009. *Jack Johnson: Reduce, reuse, recycle – 3 R song | Explore Films* [online]. Available from: https://www.youtube.com/watch?v=uSM2riAEX4U. [Accessed 1 September 2020].
Hesmondhalgh, D., 2013. *The cultural industries*. London: SAGE.
Hodgson, M., 2010. *Recycling and redesigning logos: A designer's guide to refreshing and rethinking design*. Beverly, MA: Rockport.
Jones, J., 2003. Look what we did [online]. Available from: https://www.theguardian.com/culture/2003/mar/31/artsfeatures.turnerprize2003, 31 March. [Accessed 1 September 2020].
Kaloides, M., 2006. *Don't retire ... Recycle yourself*. AuthorHouse.
Konky, G., 2015. My green school [online]. Available from: https://gali020.wixsite.com/mygreenschoolapp. [Accessed 1 September 2020].
Livingston, S., 2011. Recycling (a short story) [online]. Available from: www.amazon.co.uk. [Accessed 1 September 2020].
Makower, J., 2013. Cradle to Cradle: Joel Makower talks to Bill McDonough about 'The Upcycle' [online]. Available from: http://www.greenbiz.com/blog/2013/04/15/joel-makower-talks-bill-mcdonough-about-upcycle. [Accessed 1 September 2020].
Manchester Metropolitan University, 2014. Getting wasted [online]. www.gettingwastedgame.co.uk. [Accessed 1 September 2020].
McDonough, W. and Braungart, M., 2013. *The upcycle: Beyond sustainability – designing for abundance*. New York: Macmillan USA.
Metzger, R., 2010. Recycle: Joy Division & New Order, the Factory years [online]. Available from: https://dangerousminds.net/comments/recycle_joy_division_new_order_the_factory_years. [Accessed 1 September 2020].
NHS Blood and Transplant, 2019. Organ donation and transplantation activity report 2018/19 [online]. Available from: https://www.organdonation.nhs.uk/helping-you-to-decide/about-organ-donation/statistics-about-organ-donation/transplant-activity-report/. [Accessed 1 September 2020].
OED, 2009a. Recyclable. Oxford English Dictionary Online, third edition. Available from: https://www.oed.com/view/Entry/160129. [Accessed 1 September 2020].
OED, 2009b. Recycle. Oxford English Dictionary Online, third edition. Available from: https://www.oed.com/view/Entry/160131. [Accessed 1 September 2020].

OED, 2009c. Recycling. Oxford English Dictionary Online, third edition. Available from: https://www.oed.com/view/Entry/271857. [Accessed 1 September 2020].
OED, 2014. Upcycle. Oxford English Dictionary Online, third edition. Available from: https://www.oed.com/view/Entry/408003. [Accessed 1 September 2020].
OED, 2018. Downcycle. Oxford English Dictionary Online, third edition. Available from: https://www.oed.com/view/Entry/77097817. [Accessed 1 September 2020].
O'Hooley, B. and Tidow, H., 2010. Cold & stiff. In: *Silent June*. No Masters.
Recycle Your Organs, 2014. Recycle Your Organs [online]. Available from: https://www.facebook.com/RecycleYourOrgans. [Accessed 1 September 2020].
Rescher, N., 2000. *Process philosophy: A survey of basic issues*. Pittsburgh: University of Pittsburgh Press.
Trash & Fun, 2014. Trash & Fun [online]. Available from: https://www.facebook.com/trashfun/. [Accessed 1 September 2020].
Vanstiphout, W., 2007. Sade, Fourier, van Lieshout? In *Atelier Van Lieshout*, Allen, J., *et al.*, Rotterdam: NAi Publishers.

2

THINGS

CULTURAL PRODUCTS

All objects are value-laden, and express the culture in which they were produced. Cultural works such as films, recordings, books, images, magazines and newspapers are texts for us to read and interpret (Hesmondhalgh, 2013, p. 3). We can include in this list the artefacts created by a product designer, the useful objects such as mobile telephones with which we perform our everyday lives. In developed economies driven by consumer culture, there is no such thing as a purely functional product. Products bought to fulfil a functional purpose are almost never bought just for that. In a consumerist society, choice can be overwhelming. Products are therefore differentiated via design features, brand values, identities and logos which may bear little relation to their functional tasks, such as how well a watch allows us to tell the time. We buy and consume things for all kinds of reasons, many of which relate to cultural meaning and symbolic value. We dispose of products for the same reasons, as their meaning and value shift or lessen. Products made from materials which were previously incarnated in other objects are particularly interesting in that they inherit, visibly or not, previous meanings and symbolic values. Recycled and reused products are particularly interesting as cultural objects with a past. Some are indistinguishable from non-recycled products, hiding their remadeness. Recycled is the norm for many paper products, waste disposal sacks and aluminium food cans. Other recycled products are more overt in displaying their origin, and may tell us exactly where they came from. However, all recycled products can be viewed as cultural products, embodying values and carrying meanings that go beyond their immediate functional purpose.

ONE-OFFS

Via dedicated books, websites and television shows, we are encouraged to participate in the 'art of recycling' by making our own decorative and functional objects from everyday detritus such as plastic bottles, bottle caps and unwanted clothing. Low-cost making was traditionally popularised by children's television, but is now aimed at adults as 'upcycling', in which thrift and craftiness are presented as aspirations for the modern homemaker. Individual upcycling efforts are part of a broader trend of the return of the amateur, and a wider collective rediscovery of making, from knitting to the basic electronics of Raspberry Pi. Amateur making may give us imperfect versions of widely available and quite affordable manufactured products. It often costs more to make-it-yourself, in both time and money. But the real value lies in the process of making itself. Affluent societies which have become deskilled in a generation now rediscover the pleasures and rewards of doing-it-yourself, rather than simply buying off-the-shelf.

This collective turn towards making is clearly ideological, a response to the commodification and brandedness of so many elements of life in a developed economy. Craft-based artists and designers have embraced the recycling and reuse agenda, finding new and unforeseen applications for products that are unwanted at the end of their useful lives. These makers often invest their work with an explicit ideological commitment to raising awareness of our collective production of waste. They choose to use recycled materials over readily available non-recycled alternatives. They may also seek to revalorise low-status materials such as plastic bottles by incorporating them into products with a higher status than their original form. The most successful examples have the power to surprise us because they do not look much like discarded waste. They are made well, from materials still in their best condition. The least successful examples of crafty reuse, however, look just like waste materials re-presented in another form. Lacking utilitarian function, it may then be called art, particularly so if made by or for children. The rhetoric can be grandiose: 'We Create Art to Save the Sea!! ... We turn plastic pollution from the ocean into aesthetically powerful art – and bring dramatic attention to the problem.' Whilst a shark made from recovered marine plastic does not in itself significantly alleviate the problem of waste in the oceans, it is held to have broader symbolic value: 'These unique art pieces are part of a traveling exhibition that includes educational signage and programs that encourage reducing, refusing, reusing, repurposing and recycling' (Artula Institute for Arts and Environmental Education, 2020). Here art meets ideology.

Yet when it is done to serve an educational or instructional function, art is not assured to have any intrinsic value. Artefacts made from salvaged material can communicate good intentions, but these may not be matched by the quality of the final outcome. Earrings made by an unskilled hand from foam flip-flops retrieved from the ocean may have little more than novelty value. Paradoxically, if they are well made, the beautification of waste may have the opposite effect of celebrating the waste streams which support the creation of such outcomes. Collecting materials and making products might be meaningful to those who do it, but the results are commodities with dubious value. They are the recycled equivalent of the pointless 'crapjects' that have been the main output of the first consumer-friendly 3D printing machines (Holman, 2014). Lacking any idea of what to make using these new additive manufacturing tools, we have been happy to produce miniature models of ourselves and our favourite characters from fiction. Just as early end-consumer 3D printing was a technology in search of meaningful application, so recycled products such as goblets made from inverted chopped-off wine bottles serve mostly as articles of faith in recycling as an idea.

There is of course a long tradition of non-amateur artists using found and reclaimed objects to create assemblages that

Figure 2.1 Wine glasses made from beer bottles.

prompt us to reflect on the potential value of materials in their raw and untreated state: Marcel Duchamp's readymades; the 'combine' paintings of Robert Rauschenberg; Kurt Schwitters' use of found objects and everyday materials in abstract collage. Notable contemporary African artists reuse culturally loaded found objects in works which make powerful commentaries on current events and concerns. The Ghanaian sculptor El Anatsui uses discarded foil bottleneck wrappers in place of woven silk cloth in his piece *Man's Cloth*, a reflection on the erosion of traditional cultural values through unchecked consumerism. *Throne of Weapons* by the Mozambique artist Kester is famously made from decommissioned automatic weapons collected following the end of his country's civil war in 1992. In all these examples, the artists' considered and skilled reuse of products and materials knowingly transforms cultural value to clear purpose.

People making furniture and lighting items from materials salvaged from the aftermath of Hurricane Sandy in the US were responding to a single catastrophic event. These products were made for charitable sale, to commemorate a disaster and provide an optimistic response: 'we hope our fallen

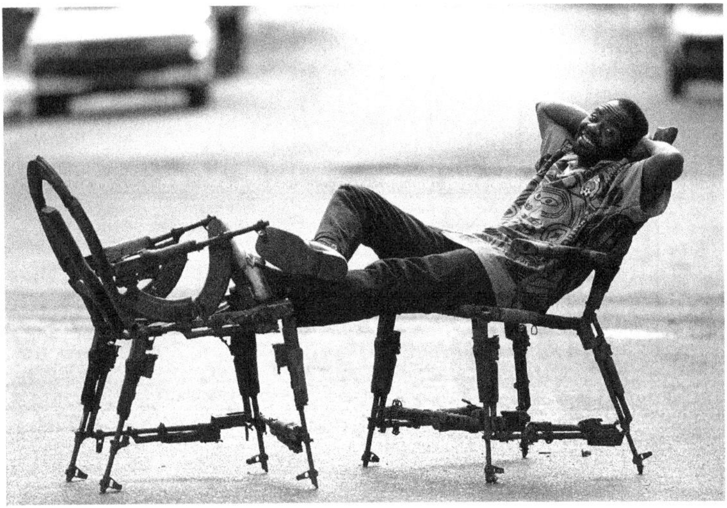

Figure 2.2 Throne of Weapons by Kester.

trees and storm-damaged building materials can be reborn as objects that represent the city's recovery' (Reclaim NYC, 2012). The storm accelerated the journey from useful product to useless material, prompting a reaction to swiftly remake things in newly useful form. These symbolically loaded pieces of furniture are like the creative reuse of recovered landmine covers and spent bullet shells – they will always carry a deeper meaning than their style and function. To own such a product is to visibly support those affected by the event which made these materials available for remaking.

Recycled products make a statement about the potential for waste to be reconstituted in a desirable form, often commodified as a consumer product. Creative reuse of waste can also be packaged for consumption as entertainment. A television series challenged three designers to turn an aeroplane, a pinnacle of human engineering, into new products, and pit their creativity against the problem of industrial waste, that inevitable by-product of industrial design (Wollaston, 2014). Sections of a decommissioned Airbus A320 aircraft, otherwise destined for disposal in landfill, were dismantled by hand and their components turned into products for the home. The participating designer-makers had strong pedigrees: one was a 'salvage designer', another an 'upcycling entrepreneur'. The exercise was introduced as a serious exploration of the potential to redeem the monetary and material value of a redundant physical asset at the end of its useful life. The plane had been superseded by a newer model, and become a victim of the relentless forces of technological innovation. We routinely replace products when an improved version becomes available, perhaps from the same manufacturer. There may be nothing wrong with the one we have – it may still function perfectly well, but is rendered inferior by an updated model. This applies to planes and mobile phones alike. Some materials within the redundant product will be salvaged, if that's economically viable, but the remainder will be discarded, their value lost. Remaking these remains, in ways that grab our attention, can make us reflect on this typical end-of-life scenario. Yet such efforts feel tokenistic. What proportion, by weight, of the Airbus A320 was actually repurposed? Such tinkering does little to address the problem of routine production of huge volumes of material waste in the form of unwanted but still perfectly serviceable products. We should look more closely at why we discard them, rather than strive to reimagine them in an inevitably lesser form.

Are there differences between artists and designers using salvaged materials in the ways just discussed? We see artworks that have clear function, and design objects that are primarily decorative. Many creative practitioners are called both artist and designer, depending on who views, or reviews, their work. It can pay dividends to be seen as both, to appeal to a broad audience. Whichever label is used, all these products of art and/or design are essentially handmade, in contrast to the mechanical mass-produced origins of many of the objects that they repurpose. This has consequences. The resulting products are invariably relatively expensive due to the high investment of labour, skilled or unskilled. Such efforts can also never offer a proportionate response to the scale of waste production. Our waste streams are generated by immense industrial manufacturing and distribution systems. Handmade responses, even when considered collectively, are unable to accommodate the huge volumes of waste generated. Upcycled designs of this kind therefore act as token gestures. Their value in responding to ever-increasing material waste streams is symbolic. At best they are examples of creative virtuosity motivated by the failings of the producers and consumers of the first-generation product. These are also mostly products of reuse, rather than recycling, if no reprocessing of materials has taken place in the making of the new object. The value of these practices does not therefore lie in a quantifiable contribution to addressing waste generation and disposal. Such practices, and their products, are propositional – they enable us to question the system of industrial production that they critique. Yet they also inescapably reside within that system, perhaps themselves participating as commercially available products. The online marketplace etsy.com has helped to dissolve any distinction between amateur and professional makers; homemade no longer means not-for-profit. Selling online allows makers of all types and levels to make money from their craft. Recycled products abound on etsy.com. Whilst this combination of factors can make us feel better, it represents a largely symbolic engagement with the deeper structural challenges of how we otherwise make things. The extent to which one-off products of reuse can have a meaningful impact, even if replicated, is likely to be limited.

Making can be empowering and seen as a form of political, economic and social activism. This is most evident in homemade objects of protest and dissent such as a mask made from a plastic bottle cut in half, tied round the face and then sealed with tape, to protect the wearer from tear gas and pepper spray. Such radical repurposing of a plastic bottle is driven by immediate functional need, rather than an ideological commitment to material reuse. We should be careful not to misread the intent of examples of reuse that are driven simply by making use of whatever materials are to hand. Gas masks are not easy to obtain. A protester's homemade version should not therefore be seen primarily as a symbol of an environmental conscience. There is a difference between recycling because we want to, and because we have no other material choice.

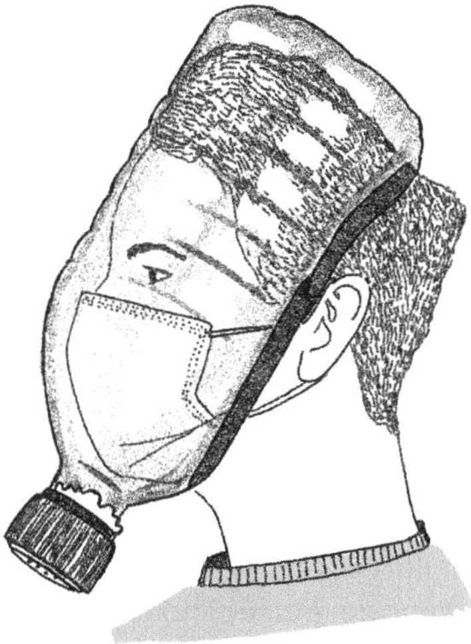

Figure 2.3 Homemade gas mask made from a plastic water bottle.

VOLUME BY-HAND (RE)PRODUCTION

Makers of one-off recycled products can increase their impact by upping the scale of what they do. Each FREITAG messenger bag is made from a unique section of a used tarpaulin cover salvaged from a truck's trailer. A carefully chosen fragment of the original commercial messaging is visible on the bag, at a scale out of all proportion to the original format. Only a cryptic link remains. A prospective owner of a FREITAG bag can even select her own section of tarp. The components of the bag are similarly made from defunct elements of other modes of transport – car seat belts, bicycle inner tubes – making the FREITAG bag an embodiment of material reuse, and a symbol of continued mobility. The materials in a FREITAG bag spent their first life on the road, which holds true for their second life in a cyclist's messenger bag.

Established in 1993, FREITAG now produces around 700,000 bags and accessories per year, and has 26 stores around the world. It uses 800 tonnes of truck tarpaulins, 150,000 car seat belts and 11,000 bicycle inner tubes per year. The product range includes 50 bags of different shapes and sizes, and 40 accessories such as wallets and sleeves for personal electronic devices (Freitag, 2020). This scale of output is not achieved by accident. Whilst the origins of the FREITAG messenger bag may have been utilitarian, the company now carefully positions itself within a particular consumer market segment. FREITAG promotes its recycled products as lifestyle accessories, the words 'detail' and 'quality' preparing us for a premium price tag. There is familiar marketing and promotional language: 'robust, water-repellent and uptown-proofed'. Less familiar is FREITAG's overt statement of their products' reused material origins. FREITAG products are made from truck tarpaulins with an average road-life of five years. The REFERENCE range presents a premium, and more costly, version made from tarpaulins which are older and more road-worn. Just as handmade leather shoes are sold using notions of tradition and heritage, FREITAG emphasises the 'workmanship' of its by-hand production methods, and reinterprets the evident wear of its reused materials by the terms 'patina', 'weathered' and 'vintage'. This language used by FREITAG knowingly responds to the common assumption that recycled products will be inferior to, and an expectation that they should therefore be cheaper than, non-recycled alternatives. Negative perceptions of quality, value and cost are hard to shake off for recycled products. The assumption is that because recycled products use secondary materials, they should be cheaper than products made from new. For most of us, second-hand or recycled means low-cost. This attitude is challenged by FREITAG, which asserts the added value of its recycled products. The company also trades on positive associations of its national identity, invoking the Swiss reputation for quality, design and functionality. Anyone sceptical of a premium recycled product is reassured by the credentials of the tradition with which the company aligns itself. FREITAG bags have been accepted into the collections of the Museum of Modern Art in New York and Zurich Museum of Design. The company is focused on material recycling, but not at the expense of other considerations. A recycled product is not excused from being functional, robust and stylish. The starting idea to make something useful from waste is applied to a material, truck tarpaulin, with undeniable but overlooked qualities. In this case, recycling undeniably adds character and value to the product. FREITAG's success – in terms of volume of products made and sold, number of employees, markets entered and financial turnover – has raised the visibility of recycled consumer products. Numerous other brands now make similar products from the same materials.

We should consider further the symbolic value of a product with such an overt recycled aesthetic. A FREITAG bag clearly communicates its material origins; it looks like a reused truck tarpaulin. The company says it selects tarpaulins primarily for their colour. They are cut to make the original branding unrecognisable, perhaps partly to avoid infringing copyright. The owner of a FREITAG bag may, however, fancy her chances of identifying the fragment which survives. The origins are rarely completely effaced. A test of a distinctive visual identity may be whether brand recognition survives the butchery. If you customise your own bag, you undoubtedly know the branding you have deconstructed. Some of each tarp's brand currency is therefore likely to be retained by a FREITAG bag. Do donors of used tarpaulins mind this reappropriation? This is a classic challenge for designers of recycled products, unsure of the risks of reusing branded visual content. Perhaps a brand owner doesn't mind you remixing their protected commercial identity. Perhaps she will prosecute for copyright infringement. How much of the meaning and value of a remade product derives from its subversion of primary branding? Does it only make decorative use of its source material? The copyright implications have never been clear.

Each FREITAG bag is unique in its appearance, if not style – there are a limited number of bag designs. This uniqueness comes naturally, deriving from the fact that each tarp has its own story, its own look and its own state of use. Each bag is cut by hand, lending it a crafted quality. Wear-and-tear from its first life on a truck gives the material character for its second life as a bag. This is a new product with a visible history, celebrating its imperfections – it is both new and not new. This contrasts with our usual expectation that new products be pristine, pure, untouched. Visible signs of wear would lead us to return other products to the store, including the sleek electronic devices that are placed inside FREITAG bags and accessories. A FREITAG bag fits a particular consumer profile with an existing ecology of products, including a single-speed racing bicycle and Apple electronic devices. The self-proclaimed individuality of a FREITAG bag and its associated lifestyle contrasts with the anonymity of the products it is destined to contain. Each iPhone is a clone of many millions of others. A new phone has a past only as materials; this is its first time as a product. It is a blank device which we try and make our own by adding protective cases, changing display settings and installing our preferred applications. Yet an iPhone never becomes individual. The meaning of a new FREITAG bag is different to that of other new products because it so clearly displays its previous life as a human-made and -used product. A FREITAG bag lives up to the stories told by the marketing of consumer electronics brands better than a smartphone ever can. It is the opposite to an anonymous output of exploitative mass-production.

An argument for recycling is that it closes a mental loop, as well as a material loop, for the consumer. If we recycle, perhaps we will act in environmentally considered ways in other parts of our lives? A recycled product is a talisman for sustainable living. A product which proudly displays its material origins is also a more powerful symbol than an anonymous recycled product, such as an aluminium food can, which keeps its previous life quiet. FREITAG successfully plays the branding game, positioning its products in relation to recognisable consumer lifestyles. It presents a mode of recycling which celebrates and derives meaning from the first lives of its materials and combines these with strong design values. FREITAG wants to distance itself from the novelty and disposability of our consumer culture, by focusing on the longevity of materials and an extension of their lives.

FREITAG uses the language of recycling, but we should recognise it as an example of material reuse. A FREITAG bag is made primarily by hand, using low-tech means. Each is made separately, to be unique. A tarpaulin undergoes no significant material transformation in its conversion to a messenger bag. This presentation of reuse as recycling is not uncommon, however, and demonstrates the challenge of distinguishing the two, discussed in Chapter 1.

An example of handmade production which does qualify as recycling rather than reuse, and in volumes greater than one-offs, is the Nordlys range of glassware by the Danish company Holmegaard. Marmalade jars and bottles are transformed into a range of handmade contemporary glasses, bowls and candle holders. As with FREITAG, each piece is unique due to the irregularity of the source material. The identification with the aurora borealis, which appear in Scandinavian night skies, is justified by the distinctive appearance of the glassware (Holmegaard, 2013). The Nordlys range is presented as upcycled – everyday food packaging transformed into decorative glassware of a higher material and symbolic value. The visual and material character of these objects is, moreover, identified with the fact that they are made from recycled glass. Nordlys glass does not look recycled in the way a FREITAG bag obviously looks reused. It lacks a clearly recycled aesthetic, as similar effects could be achieved using non-recycled glass. Nordlys' recycledness is more subtle. The manufacturer, Holmegaard, was established in 1825, so this range is an addition to a long tradition of glassmaking. This contrasts with FREITAG's much more recent founding in 1993 as a company devoted to making products from salvaged and reused materials. An ideological commitment to recycling is less dominant for Holmegaard, and the Nordlys range sits alongside the company's non-recycled standard offering. A new recycled product range can be a diversification for an existing established brand. It does not have to be a sole offering.

MATERIAL SUBSTITUTION

Many everyday products – toilet paper, refuse sacks, aluminium food and drinks cans – are commonly made from recycled materials. These products do not use recycled materials on principle, but because they perform adequately and offer good value. Recycled products manufactured at scale rarely wear their reprocessed hearts on their sleeves. Unless we are told so, we may not realise such products are in their second life. They are silently recycled, visually and aesthetically, their reincarnation evident only in their labelling. Some recycled content products are driven by principle, however, and tell us so.

The production processes of American chair manufacturer Emeco combine skilled traditional machinist labour with use of specially developed recycled materials. This minimises waste and energy use. Since its founding in 1944, the company has sought to recover and use discarded materials to make products that last. There is tension, however, between making durable products and being committed to recycling. Recycling, as an ongoing practice, requires products that do *not* endure, but readily yield their materials for new product forms. A recycling system needs regular feeding with unwanted materials. Durable products that stand the test of time, and continue to give users what they want, act against this. When Philippe Starck reimagined an existing Emeco chair he sought to 'make something so well (timeless and strong), you never have to recycle it' (van der Minne, n.d.). Product obsolescence was the enemy, and so the Hudson chair is made to last 150 years. To achieve that lifespan, it will have to meet challenges of durability, functionality and unforeseen shifts in style and taste. It is notoriously difficult to design a classic that will be valued forever, and so never need to be recycled. The Hudson is itself made from recycled and recyclable aluminium, and so whilst being against the need for recycling, it stores recyclable material for the future. The chair's intended durability reminds us that recycling should be viewed as a means of extending material lifespans, rather than as an end in itself.

Design for recycling and design for durability seem to be opposing strategies. The first treats products as temporary assemblies of materials that can be easily recovered and reused. The second seeks products that last, minimising the need for recycling. Emeco aims to combine recycling with durability, using waste materials to make chairs that will not themselves be candidates for recycling due to their enduring quality and appeal. The 111 Navy chair is a collaboration with Coca-Cola which responds to the waste stream generated by plastic drinks bottles. Lightweight bottle plastic is upcycled into a new material formula suitable for substitution into an existing chair previously made from aluminium. The familiar chair was remade in numerous colours of recycled plastic, with an inscription on the underside of the seat explaining the reinvention to the curious. 111 Navy is made entirely from post-consumer waste, namely discarded plastic bottles. Emeco's Broom chair is mostly made from pre-consumer industrial waste, namely 75% reclaimed polypropylene, 15% reclaimed wood fibre and 10% non-reclaimed glass fibre and pigment. Broom is therefore 90% reclaimed industrial waste. Where 111 Navy is a reinvention of an existing design, Broom matches new materiality with new product design. The Navy chair seeks to quantifiably address a highly visible waste stream – there are 111 bottles per chair. Yet this will clearly not significantly dent the enormous volume of Coke bottles consumed daily around the world. It would be naive to think that product manufacture which responds directly to waste streams, but without addressing the initial production of waste, can offer a real solution. Such products need recyclate in order to be made. They therefore condone rather than challenge the generation of waste. Upcycling waste into versions with higher material or functional value addresses the symptom, but not the cause, of the problem. The Broom product catalogue itself states that '250 million tons of consumer waste & 7.6 billion tons of industrial waste are generated annually in America alone' (Emeco, n.d., p. 5). More positive impact would come from directly addressing Coca-Cola's long-standing and global use of non-returnable, non-reusable plastic bottles. A recent shift to recycled content and recyclability of bottles in the Coke family of drinks brands should not distract from the fact that a one-time use business model remains, and Coke still sits firmly on the lowest rung of the waste hierarchy (Coca-Cola Great Britain, n.d.).

Manufacture of recycled chairs on the scale practised by Emeco is not enough to make a difference. Scaling up is not the answer, however. Whilst it is tempting to think that if recycled design could be done at much greater scale, we would be able to deal with current levels of waste production, we simply do not need so many remade products. We rather need to rethink the rate at which we produce waste, not just become more creative in doing something with it. Emeco's mission of 'positive de-growth' echoes Tim Jackson's (2009) 'prosperity without growth', in which consumption-based economic growth is no longer ecologically viable, and is to be replaced by a new model of prosperity which is decoupled from economic activity and material wealth. Jackson offers twelve steps to a sustainable economy, under three themes: building an economics for sustainability; enabling social flourishing; respecting ecological limits (Jackson, 2009, p. 103). In these terms, manufacturing chairs from small volumes of recovered waste has limited potential impact. It is, however, a move in the right direction. In its focus on 'making progress without making waste', Emeco does point to the need for a less consumerist system.

Emeco is a furniture manufacturer committed to using novel recycled materials. The 111 Navy chair reinvents an established product design via material substitution. Not all such examples are as inspiring. The recycled edition of Karim Rashid's Bongo stool for Offi & Company is severely compromised. The original Bongo uses Rashid's familiar design language of organic form and dazzling colours. It is both a stool and a lamp, stackable to form walls of differently coloured light. The recycled version is implacably black and does not double as a lamp. This is the poor relation of the multi-coloured originals. The manufacturer knows this when it tells us: 'you can still feel good about your purchase knowing your product is made from recycled plastic' (Offi, n.d.). This is material substitution crudely done, devaluing the product albeit with noble intentions. Emeco ensures its recycled editions have additional value via a progressive approach to material innovation. Offi has simply made a lesser version of its standard product using low-value recyclate, for which no appeal to our better natures is likely to compensate. Material substitution succeeds either when it is invisible and goes unnoticed, or when it adds value to a product. It fails when it makes a product inferior, or is perceived to do so.

A more successful recycled product designed by Karim Rashid is the bobble filtered water bottle. Use of recycled content was here integral to the product's development and a mission to create affordable products to support sustainable behaviours. The bobble is made from recycled PET plastic free of potential toxins and pollutants. Its recycled nature is core to the bobble's identity, not an awkward afterthought with little more than novelty value. Identified by its manufacturer as entirely recyclable, bobble communicates the need to retain a product's constituent material in continuing cycles of recovery and reuse. This is required if recycling is to be an ongoing activity, rather than a one-time event that only postpones the point at which materials inevitably go to landfill. Emeco proposes extreme product durability to avoid waste generation. Not all products are suited to a 150-year lifespan, however. By recognising products as temporary arrangements of materials we become more comfortable in accepting that many products quickly lose their value. In such cases we need to ensure we can salvage the value of a product's constituent materials when it reaches the end of its useful, or desired, life. Designing for recyclability is as important as using recycled material. As a desirable product that is both recycled and capable of being recycled further, bobble is more likely to advance the cause of recycling than awkward material substitutions that compromise, rather than enhance, a product.

BRANDING RECYCLED

The Freitag brothers established a start-up business dedicated to the manufacture and sale of recycled products made from materials which they salvaged themselves. FREITAG the company trades on the image of its founders – the two brothers represent the target market of young creative urbanites. EKOCYCLE is undoubtedly a higher-profile attempt to combine the power of two established international brands – beverage company Coca-Cola and media celebrity will.i.am – to promote the sale of commissioned recycled-content products. This project seeks to introduce recycling into existing mainstream consumer product brands. The language of activism ('a movement') is used to promote a consumerist environmental agenda. EKOCYCLE takes FREITAG's positioning as a desirable 'sustainable' brand to new heights, with a confidence that comes from the existing global dominance of its two partners (it incorporates the brand name 'Coke' backwards – initial products in the EKOCYCLE range got their recycled content from reprocessed plastic drinks bottles). The products in the EKOCYCLE range are 'made in part from recycled material'. Special edition Beats by Dr. Dre stereo headphones are 'made from 31% recycled materials', a fact which is presented before their comfort features and sound performance (Food & Drink Business Europe, 2013). The recycled content is achieved by material substitution in the rigid plastic headband of the headphones. This is of course an existing and well-known product with a strong brand identity. A minor degree of recycled content is bolstered by celebrity endorsement, lending the products a credibility that is perhaps inflated. EKOCYCLE's partly recycled content products benefit via association with the right names.

EKOCYCLE exploits existing brand value to create an identity for aspirational recycled consumerism. The brand and its featured products retain a conventional visual aesthetic. As with Emeco's chairs, these products do not look recycled. EKOCYCLE wants to surprise us with the unexpected origins of its branded clothing. That companies such as Patagonia have been making similar garments for years is immaterial. Now we have the power of celebrity on our side. Actors and sportspeople have long been used to sell us every kind of consumable item, from cigarettes to health insurance. Celebrities are now identified as the producers of products, owning or fronting-up the company that makes them. EKOCYCLE does not make anything, however. It is an identity that champions a range of products from other brands. It is an ecolabel for recycled products, its status deriving from its founders. Consumer interest in recycled products is encouraged by their endorsement by trusted brands with no history in this area. Ecolabels are usually bestowed by non-governmental organisations and industry associations. These are here replaced by figureheads of lifestyle brand culture. Closing the loop by buying recycled products is presented as part of a quest to make sustainable living cool. The language of activism is allied to that of desire. Even more so than FREITAG, EKOCYCLE wants to put recycled content products squarely within mainstream aspirational brand culture.

The RAW for the Oceans project is a similar union between musician Pharrell Williams and the fashion brand G-Star RAW. A range of denim clothing made from recycled ocean plastic is 'curated' by Williams, who is also identified as 'co-designer' and 'creative director' of the project. This recycled denim is presented as a material innovation offering quality and performance benefits ('up to 400% stronger') over conventional denim (G-Star RAW, Parley for the Oceans, Pharrell Williams and Bionic Yarn, n.d.). This is an improvement on crudely refashioning flip-flops salvaged from a beach into earrings, or symbolic representations of sea life masquerading as the 'art of recycling'. The further recyclability of this alternative denim, once it is no longer in fashion, is not discussed, however, inviting the familiar criticism that this simply delays the point at which materials are discarded as useless waste. The sustainability value of recycling materials that were not designed to be recycled, and cannot themselves then be further recycled, is often unclear, as will be discussed in Chapter 4. It is tempting to address the problem of waste by introducing recycled products into existing market structures. But these structures are based on ever-shortening fashion cycles that encourage ever-increasing rates of consumption. Labelling a product as recycled is dangerous if we are motivated by concerns of sustainability. Consumption and its ultimate form, consumerism, are an indisputable driver of excessive resource use. Supposedly 'green' products are always open to accusations that they do not address these deeper underlying causes and conditions of unsustainability. They may actually do harm if they mislead us into thinking we do not need to modify our patterns of consumption. If rapid, repeated and needless consumption is the problem, it is unlikely to also be the answer, whatever we consume.

PRODUCTS FOR RECYCLING

The ultimate products of recycling are those that serve recycling activity itself – plastic storage crates and wheeled upright bins, in a different colour for each type of waste. These are the essential by-products of our recycling culture, often themselves made from recycled material. Blandly utilitarian, they probably collectively contain more recycled material than the branded consumer products which seek to get us to buy recycled. Ironically, as we seek to recycle more of our waste, we need yet more of these secondary products to sort and store it. Our streets are commonly littered with plastic containers of all colours and sizes, each for a specific type of household waste. This is the new street furniture. A small front garden may not be recognisable as such because of the space demands of the required multiple bins and boxes. The type of waste that goes in each container varies from place to place, depending on how the local authority prefers to collect our paper, cardboard, glass, metal, plastic, food, textiles, batteries, etcetera. The instruments of municipal waste collection can make our pavements impassable on collection day. We may incur penalties if we fail to follow local directives on sorting our waste and offering it up for collection. Sophisticated collection schedules alternate which bins we put out each week. Our open-topped recycling crates reveal details of the lives we live privately within our homes. Via our weekly waste, we tell our neighbours what we consume, and in what quantities. Upright wheeled bins are more discreet, only disclosing their contents if we lift the lid. These brightly coloured civic servants stand rigidly like sentries, blocking access to our paths and gardens. This first link in the recycling chain can be a very public act of disclosure. It is now widely compulsory, which generates both fervour and resentment. We complain about the burden it places on us, as we diligently observe the details of the instructions we are given as to what goes where and when.

All this recycling equipment in our streets and driveways is a response to increased collection targets for household waste. Aided by our use of these objects, the collection of household waste intended for recycling rose in the UK from 11% in 2000 to 45% in 2018 (DEFRA, 2015, p. 32; 2020, p. 4). There was an EU target for the UK to recycle at least 50% of household waste by 2020 (DEFRA, 2020, p. 1). Traditionally, recycling is most effective when we segregate waste streams as early as possible. Hence the profusion of types of plastic bins and boxes that encroach on our usable outdoor space, and invade the public realm. These remind us daily of the emphasis placed on recycling, the least preferable of the 3Rs, in our thinking about waste. A system of production and consumption which prioritises waste reduction would have less need of such a blanket approach to collection for recycling. Many recycling crates are in fact quietly repurposed as general storage containers, never performing their official task. Provided for free by the local authority, they permanently reside in garages and sheds, enabling easier reuse of tools and equipment that we may otherwise lose track of. This repurposing of recycling containers

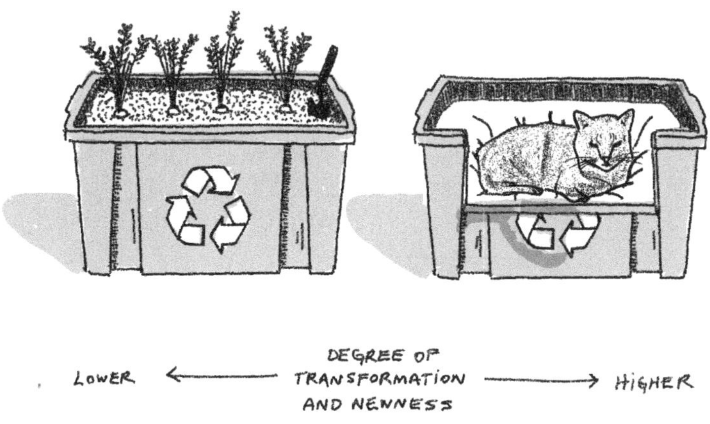

Figure 2.4 Transformations of a recycling collection crate.

as aids for reuse actually represents a step up the waste hierarchy. Recycling crates may also be used as convenient planters. Whilst this deviates from their intended purpose, it is hard to argue with as an example of adding value to the quality of our local environment.

Keen recyclers are likely to sort their waste indoors before transferring it outdoors. This has created a new category of products – recycling sorting and storage equipment suitable for use within the home. Any indoor product must of course then fit with our décor. Products that enable pre-recycling in the home need to look the part. Industrial designers who have given us ergonomic and stylish washing machines, toasters and coffee makers have applied themselves to waste separation and recycling units that integrate into the same domestic landscape. Marketed under the registered strapline 'intelligent waste',

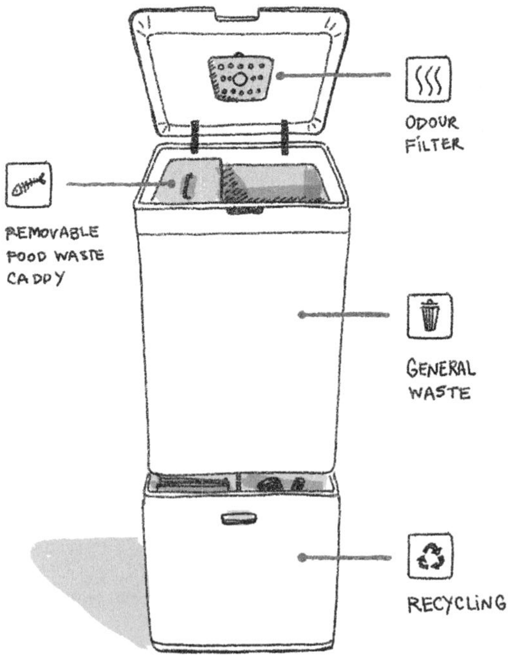

Figure 2.5 Multi-compartment domestic recycling unit.

the features of the Totem waste separation and recycling unit manufactured by Joseph Joseph include a general waste compartment 'large enough to accommodate waste from even the busiest kitchen', a multi-purpose drawer with 'easy-glide opening', a removable food waste caddy 'stylish enough to leave out on a work surface if you prefer', a fingerprint-proof stainless-steel lid with touch-button release, and a family of accompanying secondary products including replacement carbon filters to prevent unwanted odours (Joseph Joseph, 2020). Made from plastics and powder-coated steel, Totem is available in multiple sizes and colours to fit your domestic setting. It is presented in the language and digitally enhanced visual imagery of industrially designed products from smartphones to high-performance clothing. A promotional video shows the product in use in an immaculate showroom kitchen. Tasks are completed with ease whilst the object itself is fetishised. Recycling becomes another challenge to be solved by the right stylish product. We may suffer the ugliness of the bins and boxes provided by our local authority, for use outside, but we can maintain higher standards within our own walls.

RECYCLING AESTHETICS

Products that look recycled have a visual recycled aesthetic. A FREITAG messenger bag that clearly displays its material origins is different from an Emeco 111 Navy chair that is recycled so discreetly we have be told so. Recycled aesthetics do not relate only to the appearances and materialities of recycled products, however. The aesthetics of recycling as a *process* are expressed in the many activities that constitute doing recycling: the way a bottle cap is separated from a bottle, as they are rarely made from the same material; the way a recycling collection bin is hoisted onto a truck, and the noise it makes as it empties; the way a local recycling 'bring point' relates to its surroundings, for example in a supermarket car park. 'In short, the aesthetics of recycling is the aesthetics of the complex links between the life of human beings and that of their products' (Manzini, 1997, p. 41).

The aesthetics of recycling are those of the system of actions, affordances and behaviours by which materials are recovered for potential reuse. The qualities of the products which collected materials are ultimately remade as are the outcome of a long sequence of events. A recycled messenger bag is not simply an object. It is the outcome of the complex interplay of many actions, all of which are visibly or invisibly embodied within it. Recycling is a process performed by a system. Recyclability – the ability to be recycled – is therefore a property of systems, not of products. Successful recycling requires a complex supporting infrastructure. It begins with designing a product capable of being recycled, but recycling requires the involvement of a long chain of actions, human and non-human. Making a product from recyclable materials is no guarantee that it will actually be recycled. There are many points at which the chain may be broken.

We can think of a recycling system itself as a complex product which has been designed, well or badly, and can be redesigned. Products embody the properties and qualities of the systems which produce them. A recycled product is an expression of the recycling system that produced it. A recycled product tells us a story about our priorities. The type of product we prefer indicates the mode of recycling we think has most value. The aesthetics of recycling are expressed by our performance of the full range of activities associated with the recycling process, and are not just present in the visual appearance of products.

Products for recycling such as the Totem kitchen organiser position recycling within a contemporary consumer lifestyle. Far from challenging that lifestyle, it enables it. If we buy a Totem we will be inclined to use it. It may encourage us to consume more, or at least exonerate us from feeling bad about what we do consume. Every recycling bin demands to be filled, and as they proliferate we have to consume more to be able to recycle more. The aim of designers of recycling programmes is to make it as easy as possible to participate, to make it a norm. In this way recycling behaviours are encouraged for their own sake in order to meet targets imposed by governments who see increased recycling collection rates as an attainable environmental target – much more so than, for example, climate action.

Bottle Bank Arcade Machine challenges the usual aesthetics of recycling. An adapted collection bin rewards the donation of used glass bottles with flashing lights, sound effects and a points score (Volkswagen, 2009). This seeks to transform a chore into a pleasure. An abstract sense that by recycling I might enable some good to be done in the future is replaced by immediate personal reward. The simple act of putting a used bottle in a hole is transformed. Arcade Machine targets a weak point in the glass bottle recycling system – donation – with a creative approach to strengthen it. It reacts against mundane waste sorting and donation routines. Recycling collection containers must be easy to handle, convenient to store, efficient to collect, secure from pests, able to segregate waste types, durable and weather-proof, and ideally flexible enough to accommodate future potential recyclables. Efficiency and safety are the dominant considerations, not visual appeal and user stimulation. As a result, recycling has an unsatisfying aesthetic. Arcade Machine alters the aesthetics of recycling by rethinking how the bottle bank relates to its users and setting. It recommunicates recycling collection by making it analogous to something we might pay to do in a leisure park. The apparent success of Arcade Machine suggests that recycling must not only be easy to do, but a meaningful and rewarding experience.

Bottle Bank Arcade Machine redesigns the aesthetics of the first point in the recycling system – collection. Recycled content products are usually our next encounter as consumers with the aesthetics of recycling, at the end of the intervening process. The activities in between are invisible to us, demonstrating our disengagement with the complexity of the recycling system. In contrast to exemplars of clean and quiet modern industrial manufacturing such as car factories, waste recycling and reprocessing plants are noisy, dirty and smelly throwbacks to more elemental industries. This is not surprising if we consider the messy state of much of the material that we discard. Despite encouragement to segregate our waste, we dispose of much

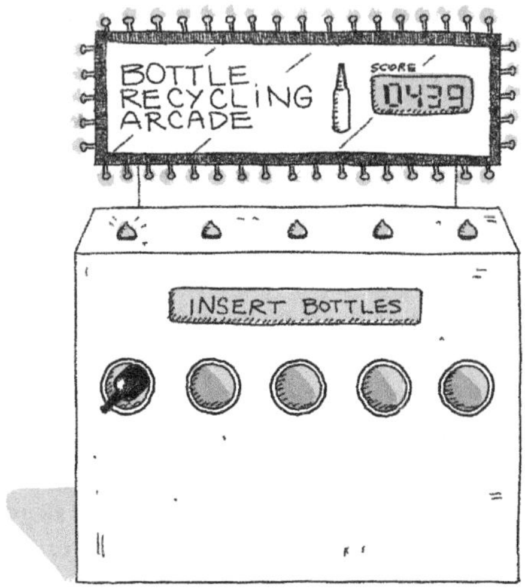

Figure 2.6 Bottle bank as arcade game.

of it with little care. Much remains to be done to sort and treat the comingled waste that results. Sorting is increasingly automated; for example, infrared cameras can detect and separate different polymers within a plastics waste stream. Much of this work is, however, still primitive and done by hand. Advanced technologies that identify and sort different types of materials can be easily disrupted by videotape or plastic carrier bags, at which point machinery must be turned off and cleared by human hands. An idealised 'reincarnation machine' which transforms discarded plastic bottles into feedstock for new products is more like resource extraction than production and assembly; '[it] is manufacturing, but it feels more like mining than making. Recycling plastic on this scale is like exploiting a natural resource' (Long, 2012).

The conventional aesthetics of large-scale recycling facilities are challenged by the proposal for Sydhavns Recycling Centre in Denmark by the Bjarke Ingels Group (BIG). Such sites are usually purely functional and located on anonymous industrial estates, hidden away from everyday life. We visit only to discard our unwanted items. The Sydhavns Centre proposal is in contrast presented within a carefully landscaped environment, built into an artificial hill which could be used as running track, picnic area or snowboarding slope as the seasons allow. The Centre is conceived as a public space with recycling facilities at its heart, presenting the hidden processes of waste management as a spectacle rather than dirty secret. The intention is to make 'back of house' recycling activities visible, so we can learn more about the materials we donate. Sydhavns is one of a series of recycling stations in Copenhagen, Denmark, where public and professionals alike can both leave their recyclables and also take away what they find (City of Copenhagen, 2018, pp. 14–19). The Centre was designed by BIG to enable visitors to easily review and scavenge the leftovers of others for direct reuse. This is forbidden at conventional waste centres, but clearly preferable in the waste hierarchy to objects being reprocessed and recycled.

The Sydhavns Centre, as conceived by BIG, attempts to demystify the front end of the waste management process. Transparent and accessible physical architecture encourages an ongoing relationship with our waste. If waste is treated as a valuable commodity at the point of collection, we may change our minds about discarding it so easily, or perhaps be more willing to accept it when it re-emerges from a recycling process in the form of a new product. If product disposal becomes more public, even ritualistic, we may rekindle our affections for products which we used to value more highly than we do now. The threat of disposal is the ultimate test of the 'emotional durability' of our relationships with our products (Chapman, 2015). Placing a pause into the recycling donation process allows us to really be sure we want to proceed. If the aesthetics of recycling are more carefully considered, we are likely to engage more mindfully in the process and understand the value, and potential limitations, of our participation.

CONCLUSION

Recycling transforms material, from one form into another, and perhaps from one product into another. This transformation is symbolic as well as physical. Recycled products mean different

Figure 2.7 A recycling centre where users exchange items.

things to non-recycled products. How and why a recycled product is made is a big part of its value. Products *of* recycling may also be products *for* recycling, if they exist primarily to enable the performance of recycling itself. A recycled content collection crate is itself an example of what may happen to the plastic bottles and containers it exists to hold. Recycled things are the final embodiment of a recycling process, and express the values of that process and the context in which it sits. Participation in a recycling system is affected by the aesthetics of that system, both technical and social. The next chapter examines the visual symbols which are used to identify recycling and its products.

REFERENCES

Artula Institute for Arts and Environmental Education, 2020. Washed Ashore [online]. Available from: http://washedashore.org. [Accessed 1 September 2020].

Chapman, J., 2015. *Emotionally durable design: Objects, experiences and empathy*. Abingdon: Routledge.

City of Copenhagen, 2018. Resource and waste management plan 2018 [online]. Available from: www.kk.dk/ressourceaffaldsplan. [Accessed 1 September 2020].

Coca Cola Great Britain, n.d. Our smartwater bottle is made from 100% recycled plastic [online]. Available from: https://www.coca-cola.co.uk/sustainability/packaging-and-recycling/introducing-our-new-smartwater-bottle-made-from-100-recycled-plastic. [Accessed 1 September 2020].

Department for Environment, Food and Rural Affairs (DEFRA), 2015. Digest of waste and resource statistics – 2015 edition [online]. Available from: https://www.gov.uk/government/statistics/digest-of-waste-and-resource-statistics-2015-edition. [Accessed 1 September 2020].

Department for Environment, Food and Rural Affairs (DEFRA), 2020. UK statistics on waste [online]. Available from: http://www.defra.gov.uk/environment/waste/. [Accessed 1 September 2020].

Emeco, n.d. *BROOM by STARCK*. Hanover, PA: Emeco.

Food & Drink Business Europe, 2013. Coca-Cola's new brand turns PET bottles into a fashion statement [online]. Available from: https://www.fdbusiness.com/coca-colas-new-brand-turns-pet-bottles-into-a-fashion-statement/. [Accessed 1 September 2020].

Freitag, 2020. Facts & figures [online]. Available from: https://www.freitag.ch/en/media/about/factsfigures. [Accessed 1 September 2020].

G-Star RAW, Parley for the Oceans, Pharrell Williams and Bionic Yarn, n.d. G-STAR x PHARRELL_RAW for the Oceans [online]. Available from: https://www.partofabiggerplan.com/work/g-star-raw-for-the-oceans. [Accessed 1 September 2020].

Hesmondhalgh, D., 2013. *The cultural industries*. London: SAGE.

Holman, W., 2014. Crapjects [online]. Available from: http://objectguerilla.com/blog/2014/10/23/crapjects. [Accessed 1 September 2020].

Holmegaard, 2013. Seeing the plate in a new light [online]. Available from: https://www.holmegaard.com/press-information/press-releases/press-releases-2013. [Accessed 1 September 2020].

Jackson, T., 2009. *Prosperity without growth? The transition to a sustainable economy*. London: Sustainable Development Commission.

Joseph Joseph, 2020. Totem [online]. Available from: https://www.josephjoseph.com/collections/totem-waste-recycling-bins. [Accessed 1 September 2020].

Long, K., 2012. Reincarnation machine: Bottle recycling factory [online]. Available from: https://www.iconeye.com/design/features/reincarnation-machine-bottle-recycling-factory-2. [Accessed 1 September 2020].

Manzini, E., 1997. The aesthetics of recycling is not in the product. In *Refuse-reuse: Making the most of what we have*, Drabbe, N., ed. Utrecht: Cultural Connections.

Offi, n.d. Bongo black stool from recycled plastic [online]. Available from: https://offi.com/products/seating/BONGO-BLACK.php?p2c=582. [Accessed 1 September 2020].

Reclaim NYC, 2012. Reclaim NYC Charity Auction [online]. Available from: https://www.facebook.com/events/547025205326124/. [Accessed 1 September 2020].

van der Minne, C., n.d. When less is more, so much more [online]. Available from: http://www.mashcollection.com/cool-business/less-is-more/. [Accessed 1 September 2020].

Volkswagen, 2009. The Fun Theory 3 – an initiative of Volkswagen: Bottle Bank Arcade Machine. Available from: https://www.youtube.com/watch?v=zCt_MzsnIUk. [Accessed 1 September 2020].

Wollaston, S., 2014. Kevin's Supersized Salvage – TV review [online]. *The Guardian* [online] https://www.theguardian.com/tv-and-radio/2014/apr/25/kevins-supersized-salvage-tv-review. [Accessed 1 September 2020].

3

SYMBOLS

OUROBOROS: A PRE-RECYCLING SYMBOL

The universal symbol of recycling shows three chasing arrows forming a triangular loop of continual motion. This was not the first symbol of circularity represented as a loop. The ouroboros is an ancient symbol depicting a snake eating its own tail in a closed circle. It originally symbolised the ancient Egyptian understanding of time as a series of repetitive cycles, evident in the flooding of the Nile and daily journey of the sun, which brought renewal after drought and darkness. Renaissance-era alchemists saw the repeating loop of the ouroboros as something to try and break, in their pursuit of an immortality based on linear time. The ouroboros is a persisting symbol of eternal cyclic renewal in many cultures and traditions, used to depict a cycle of life, death and rebirth from which we may strive to emerge. It is also seen as a fertility symbol – the tail of the snake is a phallus entering the womb of its mouth (Bekhrad, 2017). The ouroboros can therefore be considered a pre-recycling symbol, a representation of cyclical time and renewal found across many periods and cultures.

MOBIUS LOOP: THREE CHASING ARROWS

The familiar recycling symbol dates from 1970. It was created for a competition to design a symbol to appear on the recycled paperboard products of the largest paper recycler in the US. The symbol comprises three arrows, twisting and turning

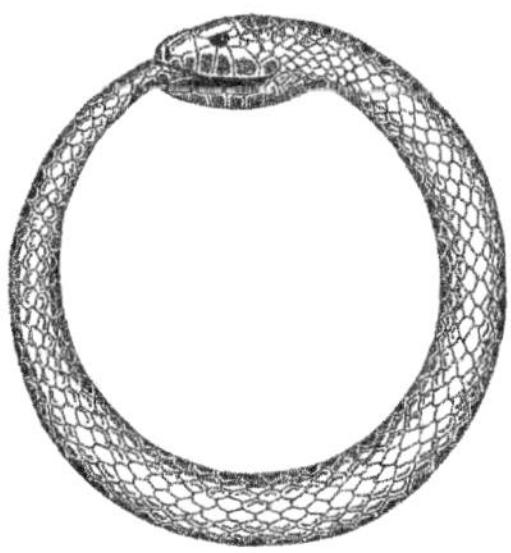

Figure 3.1 Ouroboros symbol.

as they chase each other around a central triangle of negative space. The design references August Ferdinand Mobius, a 19th-century mathematician credited with the discovery that a strip of paper twisted over and joined at the tips forms a continuous, single-edged, one-sided surface (Jones & Powell, 1999). In the words of Gary Anderson, its creator:

> The figure was designed as a Mobius strip to symbolize continuity within a finite entity. I used the arrows to give directionality to the symbol. ... I wanted to suggest both the dynamic (things are changing) and the static (it's a static equilibrium, a permanent kind of thing). (Jones & Powell, 1999, p. 2)

The winning entry was refined by the company, and licensed to paper industry groups and trade associations for widespread use with other recycled paper products. This modified version of the original design soon became the widely accepted symbol of recycling. Two useful variants followed, indicating (i) that a product contains a percentage of recycled content, and (ii) that it is itself recyclable. The symbol was rotated from the original version, with the central triangle of negative space now pointing upwards rather than downwards.

The sponsor of the competition to which the original design was submitted belatedly sought to register the recycling symbol as a trademark. This was refused due to the symbol's swift widespread currency, and so the right to its free use and modification was established from the beginning. That no one legally owns the recycling symbol, and no one can therefore

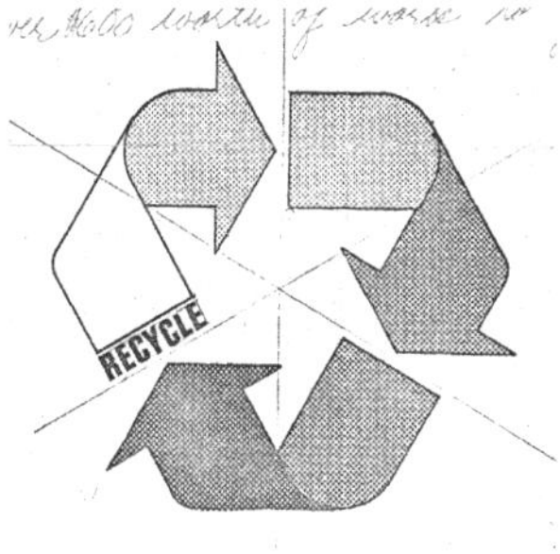

Figure 3.2 Early design of the Mobius loop recycling symbol by Gary Anderson.

Figure 3.3 Recycling symbol indicating percentage of recycled content.

Figure 3.4 Recycling symbol indicating potential recyclability.

protect or control its use, has contributed to its proliferation and diversification. Each successive appropriation and adaptation of the initial simple chasing-arrows device seeks to give it new meaning, whilst remaining recognisable as a derivation of the original. Versions of the copyright-free symbol are found everywhere from paper tags to plastic products to the logos of almost every waste management company.

Guidance exists on correct use of the symbol and its variants, in terms of both when and how it should be used, in a British Standard on self-declared environmental claims:

> The Mobius loop is a symbol in the shape of three twisted chasing arrows forming a triangle. Whenever it is used to make an environmental claim, the design shall meet the graphical requirements for ISO 7000, Symbol No. 1135. There should, however, be enough contrast so that the symbol is clear and distinguishable. Some examples of the form of the Mobius loop are provided [along with] detailed requirements concerning the use and applicability of the Mobius loop. (British Standards Institution, 2012, p. 6)

This guidance seeks to enable clarity and consistency in the environmental claims made by companies about their products. It is, however, only guidance, and not compulsory. The recycling symbol has therefore been creatively reworked by companies and organisations seeking a distinctive identity for their recycling activities. The challenge for designers of these variations on a simple theme is how to present a novel twist on an archetypal form whilst not deviating too far from the original. Any variation should still be recognisable as a recycling symbol.

The simple graphic device of three chasing arrows has become the universal signifier, the unofficial logo, of recycling. The symbol has accompanied the spread of recycling as a priority and a practice. It joins the best-known symbols in the world, including trademarked logos of multi-national organisations and corporations, religious symbols, and instructional or prohibitive icons such as the skull and crossbones. A recycling symbol has a technical function when it indicates the material composition of a product. This tells us how a product should be dealt with – including whether it can be recycled – when we are done with it. The symbol may not be trademarked, but it can certainly serve a commercial purpose when added to consumer packaging. The prominence of the symbol can indicate the degree to which a brand wants to be seen as pro-recycling, and by extension pro-environmental.

VARIANTS AND DEVIANTS

Interesting variants occur when the recycling symbol is combined with other recognised symbols and devices. It blends with Coca-Cola's iconic bottle silhouette to create a new version of a familiar logo. Disney has placed a Mickey Mouse silhouette in the central negative space of a simplified version of the classic recycling symbol, framing one of the world's most recognised characters with two stylised chasing arrows. Coke and Disney do this because an association with recycling is good for business. It is taken as a sign of being environmentally responsible. In both cases, however, a declaration of commitment to recycling seems at odds with the core principles of the company. Promoting recycling does not meaningfully offset the significant social and environmental impacts of a globally distributed and growth-based business model. Coke exists to produce and distribute sweetened and carbonated water-based drinks to as many people as possible around the world. Disney is a multi-national mass media and entertainment giant, and one of the world's biggest consumer brands. In either case, it is difficult to see how a bespoke recycling logo represents meaningful engagement with notions of sustainable resource use. We saw in Chapter 2 how Coca-Cola's EKOCYCLE brand offered lifestyle products made in part from recycled content. As noted there, a recent shift to recycled content and recyclability of bottles in the Coke family of drinks brands should not distract from the fact that a one-time use business model remains. Recycling is in this case an add-on which seeks to address one of the symptoms, whilst ignoring the cause, of the wastefulness of what the company exists to do.

The original recycling symbol's clarity and simplicity make it easily adaptable for use with existing visual communications. It is not quite so easy to introduce the principle and practice of recycling into organisations using those symbols. The recycling symbol often represents environmental tokenism, a sign of 'greenwash' by which a company wants to be seen as greener than it actually is, or could ever be. Because use of the recycling symbol is not regulated, and recommended codes for its use are voluntary, consumers are at the mercy of anyone who uses the symbol. What a particular recycling symbol really means can be very unclear.

As we saw above, the origins of the recycling symbol lie in the labelling of recycled paperboard products. The currency of the symbol now goes way beyond such simple identification of materials. Like the Nike Swoosh or McDonald's Golden Arches, the recycling symbol does not need words. It stands proudly on packaging around the world, seemingly requiring no additional explanation or translation. But easy recognition is not the same as proper understanding. We immediately recognise the device and know that the chasing-arrows symbol is meant to convey, in a broad sense, recycling. What this means for a particular product in a particular place is often unclear. Gary Anderson, designer of the original symbol, considers its widespread appropriation and adaptation a good thing – the more variations the better (Jones & Powell, 1999, p. 2). Yet, however it is used, the symbol should provide clear guidance of some form – it should mean something specific, and direct us to do something useful. The proliferation of so many variants of the recycling symbol can obscure useful nuances of meaning. Specific intentions behind uses of particular recycling symbols can be lost. Visitors to one of Britain's most popular news websites were asked to decode the meanings of a number of widely used recycling symbols (BBC News, 2007). The data generated was not recorded, although your own results could be viewed on completing the quiz. The fact that the topic was featured in this way in the magazine section of such a popular website indicates that it addressed a common challenge. The quiz appeared in response to a published survey on packaging recyclability. Its introduction referred to consumers' confusion at trying to make sense of the numerous recycling symbols commonly used on retail packaging in the UK. Questions asked for the correct meaning of ten recycling symbols, including the Green Dot logo and symbols for various plastics, metals and compostable materials. The symbols have a range of intentions and purposes, such as: material composition – including recycled content; recommended disposal actions – for example a bottle bank for glass; participation by the product manufacturer in a broader recycling programme, at either national or international level. These labels are useful to the extent that they aid effective recycling practice. Recyclability is a necessary, but not sufficient, precondition for recycling. The fact that a particular type of plastic is in principle recyclable (most things are) does not mean it will be recycled. Many of these symbols are simply reminders of what to do with a piece of packaging when we are soon done with it. In these cases, the recycling instruction equates to correct disposal by the consumer.

Collection of waste and its separation into appropriate waste streams form the first necessary step in the recycling process. Many further steps must follow this initial act for a bottle or jar to actually be reprocessed into a new product and reused. Yet most considerations of recycling focus only on what we immediately do with our consumer waste. In dropping a bottle into a bottle bank we are simply feeding it into the front end of a complex and distributed recycling system. There are many points at which that system can fail – collection, sorting, transportation, reprocessing and reuse – in which case the value of the recycling effort is limited. Effective recycling, in this holistic sense, requires that lots of things are done well along the material reprocessing path. This can vary greatly due to local circumstances. One local authority may collect materials for recycling which a neighbouring authority does not, due to having different materials handling capabilities and reprocessing facilities. This explains the unconfident message on consumer packaging: 'check local recycling'. The On-Pack Recycling Label (OPRL) scheme seeks to provide clear advice on how we can recycle, acknowledging the varying provision of recycling facilities across the UK (OPRL, n.d.a). Recyclability is not a given, but always depends on local facilities. As consumers we require this additional local information, to take part in whatever recycling programmes are available where we are. The original On-Pack Recycling Label scheme had three categories: 'widely recycled', 'check local recycling' and 'not currently recycled'. Each had a technical definition: 'widely recycled' meant 75% or more of councils provide household recycling collection facilities for that packaging type; 'check local recycling' denoted 20–75% of councils did so; 'not currently recycled' meant fewer than 20%. The last category did not mean that no local facilities exist anywhere, only that this was probably unlikely where you are (Recycle Now, n.d.). The scheme's attempt to simplify complexity could, however, mislead us into thinking that some materials, for example plastic film, are not recyclable anywhere, when in fact they may well be where you are. The three-part labelling system has now been replaced by a binary labelling system: 'Recycle' or 'Don't Recycle'. This responds to public confusion with the previous version, and apparent desire for an immediate Yes/No decision when looking at on-pack labels for recyclability. 'Recycle' indicates that 75% or more of UK local authorities collect that type of packaging, and it is then effectively sorted, processed and sold as recyclate for use in new packaging or products. 'Don't Recycle' indicates that fewer than 50% of UK local authorities collect that type of packaging and/or it is not effectively sorted, processed or sold as recyclate for use in new packaging or products. This distinction is intended to reduce contamination of recyclables by non-recyclables at the collection stage (OPRL, n.d.b). However, the range inbetween these two figures, i.e. 50%–75%, is not clearly discussed. A positive 'Recycle' label also still requires that we know where to put a particular material for recycling, and have the will to do so. The revised scheme has edited the options available to us when faced with the dilemma of what to do with an item of packaging waste, by presenting a deceptively simple answer to the complex question: should I recycle this? The On-Pack Recycling Label scheme is intended to supplement specific and updated information given online, which would perhaps help us to make an informed decision. The labels incorporate the circular symbol of 'Recycle Now', the national recycling campaign for England, supported and funded by the UK government and used by the majority of English local authorities. The Recycle Now device is reminiscent of the Mobius symbol, but sufficiently distinctive to have its own identity as the symbol of a specific national recycling campaign. It seeks to brand recycling collection in the UK. It could of course be adapted for use elsewhere if locally specific information were available to inform the definitions of each category of recyclability.

The online quiz discussed above reveals the complexities in consumer product labelling relating to recycling. Recycled content is quite different from recyclability, but they may be found together. A plastic water bottle may be recyclable, made from recycled material, or both. We will be more inclined to donate a product for recycling if it carries a label directing us to do so. Even better if we see the materials collected for recycling embodied in new products. Scepticism over what happens to our waste once it has been collected is best addressed by showing us how it may get a new lease of life. Symbols showing recycled content are less common, however. We should also distinguish recycled content that comes from post-consumer waste, i.e. from us, from manufacturing waste that is fed back into production before it gets anywhere near a consumer. Post-consumer waste has had a more eventful journey than waste that remains in a factory. It is useful to tell us this when we look at a recycled content product.

The European Green Dot logo appears on much consumer product packaging, in Europe and beyond. This logo indicates that the company placing a product on the market contributes to the collective costs of material recovery and recycling in that particular product or material sector. It does not mean that a particular package bearing the label has either been recycled or is itself recyclable. The Green Dot signifies that for each piece of packaging, a financial contribution has been paid to a national packaging recovery organisation (Valpak, 2020). The symbol is used internationally. In countries where the Green Dot scheme applies, such as Germany, producers use the logo to signal to consumers that they participate in a compliance scheme, and contribute to the financing of collection, sorting and recovery of packaging. Manufacturers may, however, also use the Green Dot symbol in countries where the scheme does not apply, such as the UK. The UK has its own Packaging Waste Recovery Note system to support the recovery and recycling of packaging waste. This involves the purchase of certificates of recycling evidence to show that a company has met their obligations to fund the recycling and recovery of packaging waste, as required by UK Packaging

Waste Regulations. If a company exports its products to other European countries, it may well need to also display the Green Dot on its packaging to demonstrate compliance with that scheme elsewhere. The company may therefore purchase a Green Dot sub-licence in the UK to avoid printing two sets of packaging – one without the Green Dot for the UK, and one with the Green Dot for exports. The Green Dot then appears on packaging sold in the UK despite that fact that the scheme does not apply there. This is unlikely to be understood by consumers, who will naturally assume the Green Dot is an assurance of participation in a UK recycling scheme.

We increasingly talk about non-recyclables, rather than waste. Like all negatives, non-recyclability is a deviation from a positive norm, in this case recyclability. Recycling is the expectation, albeit one more honoured in the breach than the observance. A sophisticated system of recycling symbols is fine in principle, but if it is not followed in practice, little is achieved. The more complex the recycling symbol system and its categories, the greater the potential for confusion. If we are unable to make a clear connection between an object in our hand, its labelling and the available recycling options, the outcome is likely to be indiscriminate disposal in any available receptacle.

Figure 3.5 Symbol for non-recyclable 'other waste'.

Thus food-contaminated paper and wet single-use coffee cups are mixed with used copier paper, and greasy pizza boxes are mixed with clean cardboard packaging. Such errors make recyclables unrecyclable, undoing the good work of those who were more observant. An entire batch of recyclate can be ruined by one item of contaminated waste.

An ecolabel is a badge or symbol intended to reassure us of the environmental credentials of a product or piece of packaging. A Mobius loop formed of chasing arrows is the archetypal ecolabel. The recycling symbol, usually green in colour, has become the unofficial logo not just of recycling but of environmentalism itself. The same thing has happened to the symbol of the Campaign for Nuclear Disarmament, which has become a general-purpose peace symbol beyond the anti-nuclear movement, similarly aided by lack of copyright protection. The recycling symbol is used when a general pro-environmental symbol is required. A UK bank encouraged its customers to switch to electronic account statements using a visual icon which wrapped the Mobius symbol around a representation of a printed document. The bank wants us to go online to view our statements, rather than receive printed documents in the post. In doing so we are not actually recycling anything, but the recycling symbol is nevertheless used to communicate what is presented as a pro-environmental option. In going paperless we actually go higher up the waste management hierarchy, by reducing the initial production of materials and inevitable generation of waste. This is better environmentally than recycling yet is represented by a recycling symbol. A relatively weak pro-environmental strategy (recycling) therefore stands in visually for a stronger one (waste reduction via dematerialisation), and an opportunity to distinguish between levels of the waste hierarchy is lost. Use of the recycling symbol here undersells what the bank is doing environmentally. The fact that recycling symbols are everywhere, and not always used with care, as in this case, prevents us from considering recycling critically in relation to broader notions of sustainability.

Variations on the classic chasing arrows rarely deviate from the archetypal form, otherwise we wouldn't recognise them as recycling symbols. Disney and Coca-Cola versions both use

Figure 3.6 Official recycling symbol used in Taiwan.

two rather than three arrows, but are easily recognisable as recycling symbols. A variant which takes greater liberties with the form of the original symbol is used in Taiwan. Four arrows represent a national recycling programme, each identifying one of four stakeholders: (i) residents who deposit materials for recycling, (ii) recycling industries who buy the collected materials, (iii) local authorities who organise waste collection, (iv) a national recycling fund which subsidises collection of recyclables via a levy on manufacturers and importers of products which enter the waste stream (US EPA, 2012; EPAT, n.d.). This fourth element constitutes an extended producer responsibility scheme similar to the Green Dot discussed above. This symbol visually represents, and reinforces, a coherent national strategy for effective material recycling. The four green arrows of the stakeholders point to the centre of the symbol. The negative space between them forms a further four arrows which point outwards, symbolising the progressive system which is credited with raising the national recycling rate to one of the highest in the world (Middlehurst, 2019).

The three chasing-arrows recycling symbol has also been used in the Good Life Goals, a restatement of the technical framework of the United Nations Sustainable Development Goals (https://sdgs.un.org/goals) in the form of relatable and achievable actions (SDGBusinessHub, 2020). The Good Life Goals seek to enable us to put the SDGs into practice as individuals. They use the visual language of emojis to personalise and humanise the SDGs, converting them into

Figure 3.7 Good Life Goals' 'Live Better' emoji.

behaviours linked to activities, products, services and lifestyles, not expressed directly by the SDGs themselves. The Good Life Goals are carefully aligned with the 169 targets and indicators of the Sustainable Development Goals. SDG 12 – 'Responsible Consumption and Production' – becomes the Good Life Goal 'Live Better', and contains five actions: (1) Learn about sustainable development; (2) Reuse, repair, recycle, share and borrow; (3) Waste less food and use leftovers; (4) Collect friends and experiences, not just things; (5) Demand that businesses respect people and planet. The 'Live Better' emoji adapts the classic three chasing arrows recycling symbol, exploiting its widespread recognition across cultures. The emojis are available in several languages as well as English, the accompanying text changing in each case whilst the images remain unchanged. The 'Live Better' emoji is perhaps the most recognisable as a call to pro-environmental action, due to its use of the familiar symbol. Its visual appeal is perhaps lessened a little by the fact that it is coloured in shades of brown, in contrast to the brighter colours of most of the other emojis, but the fact is that its form is probably enough for it to be quickly understood. The five actions of 'Live Better' are also things we are perhaps already doing, or at least are aware that we could be doing in our existing lives. Crucially, however, 'Live Better' and its actions, including recycling, are just one of 17 Good Life Goals. Sustainable living requires that we address all 17, not just one.

CULTURAL USE

The Mobius loop has been widely adopted as a symbol not just of recycling but of broader environmentalism. The symbol's simplicity means that just about anyone can produce their own version without losing the original meaning. As already discussed, this is aided by the fact that no one owns copyright for the symbol. As a result, it appears almost everywhere there is graphic design. On product packaging, the symbol seeks to inform us of potential recyclability, or recycled content. The symbol is also used much more broadly as a badge for a lifestyle in which recycling plays a key role.

T-shirts and hoodies bearing the recycling symbol are widely available, a declaration of faith in a hopeful future. If a T-shirt which puts the recycling symbol on your chest is not enough, you can get a tattoo. A T-shirt is easily taken off, perhaps only worn once. A tattoo makes a more permanent declaration. If a slogan or symbol has enduring meaning to you, why not imprint it directly onto your skin? *Tattoo* by British artist Gavin Turk is a photograph of a pair of upturned hands with a fine black outline of the classic recycling symbol apparently tattooed on each palm. The artist's website provides a number of possible references for the image: the cycle of life and death; the stigmata produced by the nails through the hands of Christ; the artist as a recycler of images (Turk, n.d.). This range of potential metaphorical meanings, including resurrection, shows the symbol's versatility, and is further proof of the wider significance of recycling as an idea, as discussed in Chapter 1.

A Mobius tattoo is available digitally to players of the life simulation video game *The Sims*, in which virtual people populate pre-constructed or player-built environments. We give our Sims personality traits, and direct them as they seek to satisfy their desires (Electronic Arts Inc., 2020). Additional user-created features allow us to customise the appearance of our Sims; these include a recycling symbol tattoo, available in several sizes within the game's 'tattoo tool' (The Sims Resource, 2010). The availability of the Mobius symbol in the Sims' virtual environment shows its currency. The symbol is, however, functionally redundant in this context, as Sims gameplay is virtual

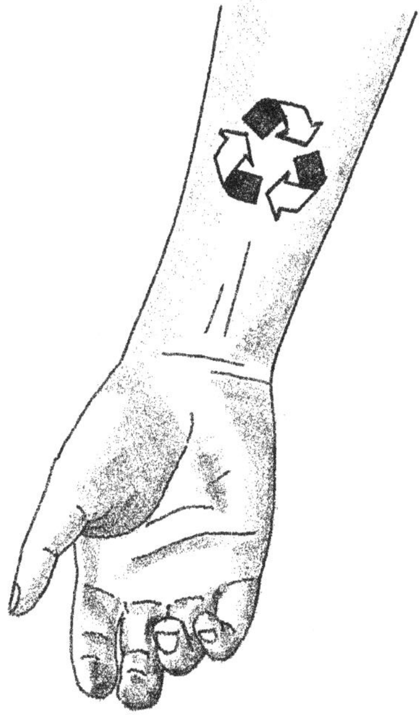

Figure 3.8 Mobius loop tattoo.

and entirely disembodied. No physical materials are used or wasted in the construction of the gaming environment. In this digital domain the Mobius loop is purely symbolic. There is no actual material recycling to be done in the world of the Sims.

Recycling also features in the play world of the classic plastic construction toy LEGO. The Recycling Truck set in the LEGO City range is aimed at children aged 5–12 years. It includes: a truck; a forklift vehicle; three recycling bins in different colours for paper, metal and glass; accessories such as a wheelbarrow, shovel and brush; two workers, and a member of the public keen to recycle. Like the many books encouraging children to recycle, this LEGO set is ideologically loaded. Recycling is presented as an uncontested desirable social norm. A recycling bin also features in another set in the LEGO City range. The Fire Motorcycle set includes: a fire-fighter minifigure with fire extinguisher; a fire motorcycle; a large recycling container in the colour green and with the classic Mobius loop symbol; and two large flames. The scenario of the set is that the recycling container is on fire and must be extinguished by the firefighter, with the assistance of the child playing with the set. Whilst the Recycling Truck set is all about recycling, the recycling container in the Fire Motorcycle set seems incidental to the main fire-fighting theme. Yet a public-use recycling container is a likely site for a fire. A large volume of recyclables in a recycling container can constitute a fire hazard. Seen as a security risk, waste bins are sometimes removed from busy public spaces. Recycling bins are an equally credible site for a fire, either accidental or deliberate. In presenting these two qualities of the recycling system in LEGO world – enthusiastic participation in collection of recyclables, and the risk of them going up in flames – LEGO reflects some of our own ambivalence to the aesthetics of recycling. The clarity of the idea of recycling is often not met by the messiness of the practice. LEGO's focus on recycling is not manifest in recyclability of its bricks. The company encourages reuse before recycling of its sets and their parts, launching its own takeback and redistribution scheme (LEGO, 2020a). Local recycling schemes should be investigated to see if they might accept unwanted or broken plastic bricks only when all options for redistribution and reuse have been considered. As a large plastic products manufacturer, LEGO seems to understand the value of the waste management hierarchy. The company remains committed to product durability to enable long-term use and reuse, even as it seeks to replace its traditional oil-based plastic with a plant-based alternative (LEGO, 2020b).

The Mobius loop symbol lends itself equally well to corporate and non-corporate adaptation. Coca-Cola and Disney have integrated the symbol with their existing graphic languages, adding a familiar character silhouette to the centre of their versions. Guitar silhouettes replace the conventional chasing arrows in a symbol used at a live music festival, although the broken guitar necks at the tips of the triangle are in clear need of repair.

The Mobius symbol can even acquire religious symbolism. By rotating the symbol and adapting the arrows, the central negative space can form the six points of the Star of David, symbol of Judaism and Jewish identity. The Star is composed of two overlaid equally sized equilateral triangles. Subtle alteration of the conventional form of the recycling symbol accentuates an outline shape that is almost present in the original. This reorientation of the conventional symbol returns to Gary Anderson's intention, later modified to the familiar version (see Figures 3.2–3.4). As with the Coca-Cola recycling symbol, the Mobius is here reworked to incorporate another well-known graphic device, thereby associating the recycling agenda with another ideology. An association with recycling is likely to be good for a business and a religious group alike. The Star of David appears centrally on the national flag of Israel. This recycling symbol therefore communicates a sense of place geographically, culturally and politically.

Figure 3.9 Guitar-based variation of the recycling symbol.

Figure 3.10 Recycling symbol incorporating the Star of David.

Figure 3.11 US federal government recycling logo with eagle.

Creative use of the central negative space in the recycling symbol is also made by the US federal government recycling logo featuring a bald eagle. This combination of the chasing arrows with a well-known national symbol taken from the non-human world promotes a commitment to recycling as a core element of national identity. This is not yet matched by practice, however; 34% of household waste in the US was recycled or composted in 2015, suggesting the symbol represents ambition rather than achievement (US EPA, 2018, p. 2).

CONCLUSION

The Mobius loop recycling symbol is used variously by companies, organisations and individuals. From its origins as a technical symbol for use by the paper reprocessing industry, it has become a badge of broader cultural identification. The three chasing arrows form one of the most recognised and adapted symbols of our time. The early failure to protect usage of the symbol has allowed it to be widely reproduced and reinterpreted. Manufacturers of consumer products have their own versions of the familiar device, marrying it to their own existing visual identities. This proliferation has brought considerable drift away from the intended meaning of the original version created in 1970 by Gary Anderson. Variants of the classic symbol often do no more than suggest that we are participating in recycling in some way. This may make us feel better about our consumption of resources, just as we may seek to offset our feelings of guilt over the carbon emissions arising from an airline flight. Such feelings may, however, be little more than personal greenwash. The fact that something can in principle be recycled, and carries a logo that says so, does not mean that it necessarily will be, even if we do our part and donate it to the first stage of the extended recycling process. Many products are not fit to be recycled, such is the cost in energy and other resources of seeking to reprocess a product which was not designed with that end in mind. The extended family of graphic devices deriving from Gary Anderson's original design collectively represent the contemporary brand identity not just of recycling, but, by extension, of all that is green and sustainable in intent. The examples of non-technical use of the Mobius loop symbol considered here demonstrate its symbolic value beyond the requirements of specialist waste management. At times, the familiar symbol becomes an empty decorative device. Adding the symbol to a computer mouse pad to indicate our commitment to recycling makes the basic mistake of creating an unnecessary product to make a point about conservation of resources. Many recycled content products fall into this category. They have novelty value, and perhaps raise awareness, but fail to demonstrate recycling as a constructive activity. The recycling logo has clearly exceeded its initial remit. The next chapter critiques the privileged position of recycling at the heart of the sustainability agenda.

REFERENCES

BBC News, 2007. Recycling symbols quiz [online]. Available from: http://news.bbc.co.uk/1/hi/magazine/7057882.stm. [Accessed 1 September 2020].

Bekhrad, J., 2017. The ancient symbol that spanned millennia [online]. Available from: https://www.bbc.com/culture/article/20171204-the-ancient-symbol-that-spanned-millennia. [Accessed 1 September 2020].

British Standards Institution, 2012. BS EN ISO 14021:2001+A1:2011. *Environmental labels and declarations – self-declared environmental claims (Type II environmental labelling).*

Electronic Arts Inc., 2020. *The Sims* [online]. Available from: https://www.ea.com/en-gb/games/the-sims. [Accessed 1 September 2020].

Environmental Protection Administration Taiwan (EPAT), n.d. Recycling system: 4.1 the four-in-one resource recycling system [online]. Available

from: https://recycle.epa.gov.tw/en/recycling_knowledge_01.html. [Accessed 1 September 2020].
Jones, P., and Powell, J., 1999. Gary Anderson has been found! *Resource Recycling*, May, pp. 1–2.
LEGO, 2020a. Replay [online]. Available from: https://www.lego.com/en-us/campaigns/replay. [Accessed 1 September 2020].
LEGO, 2020b. Sustainable materials [online]. Available from: https://www.lego.com/en-gb/aboutus/sustainable-materials. [Accessed 1 September 2020].
Middlehurst, C., 2019. How the country once nicknamed 'Garbage Island' cut waste by 30% [online]. Available from: https://bit.ly/3ixLZHU. [Accessed 1 September 2020].
OPRL, n.d.a. On-pack recycling label [online]. Available from: https://www.oprl.org.uk. [Accessed 1 September 2020].
OPRL, n.d.b. What is the scheme? [online]. Available from: https://www.oprl.org.uk/get-involved/what-is-the-scheme/. [Accessed 1 September 2020].
Recycle Now, n.d. Recycling symbols explained [online]. Available from: https://www.recyclenow.com/recycling-knowledge/packaging-symbols-explained. [Accessed 1 September 2020].
SGDBusinessHub, 2020. Good Life Goals [online]. Available from: https://sdghub.com/goodlifegoals/. [Accessed 1 September 2020].
The Sims Resource, 2010. Recycling tattoo [online]. Available from: http://www.thesimsresource.com/artists/tenshiak/downloads/details/category/sims3-makeup-costumemakeup-facepainting/title/recycling-tattoo/id/1008947/. [Accessed 1 September 2020].
Turk, G., n.d. Tattoo. Available from: http://gavinturk.com/artworks/image/33/. [Accessed 1 September 2020].
United States Environmental Protection Agency (US EPA), 2012. Recycling regulations in Taiwan and the 4-in-1 Recycling Program. Workshop materials on WEEE management in Taiwan, Handout 1 [online]. Available from: https://www.epa.gov/sites/production/files/2014-05/documents/handout-1a-regulations.pdf. [Accessed 1 September 2020].
United States Environmental Protection Agency (US EPA), 2018. *Advancing sustainable materials management: 2015 tables and figures. Assessing trends in material generation, recycling, composting, combustion with energy recovery and landfilling in the United States*. Washington, DC: US EPA.
Valpak, 2020. The Green Dot symbol [online]. Available from: https://www.valpak.co.uk/beyond-compliance/green-dot-licence#.UPAGi_IjP68. [Accessed 1 September 2020].

4

SUSTAINABILITY

THE BENEFITS OF RECYCLING

Recycling has enjoyed a privileged position at the heart of what we consider to be 'sustainable' thinking and practice:

> Recycling is widely considered to be positive for the environment. People instinctively believe that re-using materials from products which might otherwise end up in a landfill site must be environmentally beneficial. The idea that recycling is intrinsically 'green' is promoted widely – by politicians, local authorities, manufacturers, journalists and, indeed, most environmentalists. It has come to symbolise good environmental practice. (Cooper, 1994, front cover)

This summary still holds true. It therefore seems strange to ask, how sustainable *is* recycling? We take it for granted that we should try to recycle wherever we can, and that the challenge lies in knowing what we can recycle, and how. It is, however, useful to question the fundamental assumption that recycling is always a positive step towards sustainability. The waste hierarchy of the 3Rs states that we should first strive to reduce our production of waste, then reuse what we do produce and then recycle what remains. A thoughtful approach to managing material waste requires more than a singular commitment to recycling, but that is what we default to. By ignoring the higher levels of the waste hierarchy, our obsession with recycling may be a driver of *un*sustainability, as we focus enthusiastically on the wrong thing. This seems perverse to anyone engaged with the popular discourse around sustainability. Received opinion holds that recycling is a pillar of environmentalism;

to challenge that would be heresy. How can recycling be bad, when it is such a potent symbol of our desire to manage resources more efficiently? The danger lies in focusing too narrowly on the least effective approach to recovering the value of the things we use. Recycling is no more than environmental tokenism if we fail to address the underlying unsustainability of our system of globally distributed production and consumption. By intervening at the final stage in the flow of materials through this system, recycling can distract us from the necessary holistic consideration of sustainability. Recycling can stop us thinking about the bigger picture, giving us false reassurance that we are doing enough. Recycled products made from waste are symbols of our good environmental intentions, as discussed in Chapter 2. We should, however, consider very carefully the sustainability of those products, and the systems which produce, deliver and manage them. We need to examine the extent to which recycling makes a meaningful positive contribution to the problem of waste.

Recycling has long been central to the modern environmental agenda. To be 'green' is to recycle. Recycling has become a shorthand for sustainability itself, and is commonly viewed uncritically as an absolute good in and of itself. Recycling is usually presented to us as being an unproblematic, even incontestable, element of green behaviour. The online open innovation platform OpenIDEO encourages people across the world to collaborate to develop solutions to social challenges. OpenIDEO publishes a challenge, to which the online community responds. One such challenge, seeking ideas to enable better recycling habits at home, presented the common positive view of recycling:

> When we recycle we conserve precious natural resources that are often in finite supply. Recycling allows used materials to be converted into new products without any extra raw materials being used. It can also allow us to reuse materials again and again. ... Recycling is something in which we all have a role to play. It's one of the easiest ways we can contribute to protecting our environment. (OpenIDEO, n.d.)

Here and elsewhere, recycling is directly linked to environmentalism and sustainability.

Recycling is commonly seen as having a number of environmental benefits (Ackerman, 1997, p. 21)

BENEFIT #1 RECYCLING DIVERTS WASTE FROM LANDFILL

As the third level of the waste hierarchy, recycling sits just above burying materials in the ground. Landfill represents the out-of-sight approach to dealing with our waste, and its common use has allowed us to forget about the inevitable material consequences of our consumerist lifestyles. Use of landfill demonstrates a disregard for the value of waste. Decomposition of materials in landfill sites is also a significant source of methane emissions, a much more powerful greenhouse gas than carbon dioxide; municipal solid waste landfills accounted for 15% of human-related methane emissions in the United States in 2018 (US EPA, 2020). When we successfully recycle, materials are diverted from the vast waste streams flowing into landfills, reprocessed and made into new products for our reconsumption. If this happens perpetually, nothing will ever go to landfill, which is the goal of a circular materials economy. Realistically, most recycling currently gives one more life to materials before they fulfil their destiny as landfill. A pencil made from the coffee cups delivered by vending machines offers only a short-term rescue for polystyrene. It will now reach landfill as shavings from a pencil sharpener, rather than a once-used cup. In the worst case, the recycled plastic pencil will sit unused on a desk – an impotent symbol of salvaged material value, let down by the fact it does not work as well as a conventional pencil. In this case, waste has been diverted from landfill, only to be stored above ground in a new but unused product form which required additional resources to produce.

BENEFIT #2 RECYCLING CONSERVES RESOURCES

When we successfully recycle, used materials are converted into new products, reducing the need to consume additional resources. Recycling therefore reduces the need for extracting (mining, quarrying and logging), refining and processing raw materials, which often cause significant air, land and water pollution. Recycling can provide industry-ready materials much more efficiently, as in the case of extracting precious metals from discarded mobile phones. If used materials are not recycled, new products are made by extracting fresh, raw materials from the Earth's finite systems. This take-make-waste model is the legacy of the Industrial Revolution, and its core belief that progress was served by maximising the throughput of resources in a linear system which converts useful resources into useless waste. We now realise that we live on a finite planet, and cannot exploit its resources without thinking of their conservation and replenishment. In slowing down the rate at which new materials flow through this system, and seeking to make it circular rather than linear, recycling offers a way out of the endgame of rampant material consumption.

BENEFIT #3 RECYCLING REDUCES ENERGY USE

Making products using recycled materials typically consumes less energy than using new materials, even when we factor in associated costs such as transportation between the numerous facilities of a typical recycling system. Energy is needed to extract, refine, transport and process raw materials ready for use by industry. Energy use equates directly to harmful environmental emissions in most parts of our global fossil fuel-based economy. Minimising energy consumption is therefore a valuable tactic in our efforts to address climate change.

These three benefits of recycling relate to waste management and the use of recyclate instead of new raw materials. They motivate us to do all we can to feed the recycling system with waste generated by our consumer lifestyles. As consumers we are, however, at the end of the line in terms of scope for positive impact. Production waste dwarfs consumer waste:

> the waste coming out of our houses is just the tip of the iceberg. For every one garbage can of waste you put out on the curb, 70 garbage cans of waste were made upstream just to make the junk in that one garbage can you put out on the curb. So even if we could recycle 100 percent of the waste coming out of our households, it doesn't get to the core of the problem. (Young & Sachs, 1994, p. 13)

The emphasis placed on household recycling by national and local governments, facilitated by increasingly sophisticated collection schemes, suddenly seems inadequate. The statistic given above directs us to focus instead on efficiencies of production. The detritus of consumption, in the form of household waste, is certainly part of the same extended process of material extraction, transformation and use. The scope for reduction of the impacts of that process is, however, greater the further upstream we go towards the source. We should therefore focus not just on the design of consumed end-products, but on the most elementary aspects of their production.

The three reasons to recycle given above are expressions of one essential benefit: 'the advantage of recycling is that it leads to less stuff: less waste in disposal, or less virgin material use in production' (Ackerman, 1997, p. 23). The value of recycling lies in the extent to which it reduces the volume of material we handle. The most effective way to do this is by waste reduction – either producing less stuff in the first place, or minimising material use, for example by making products lighter (thereby also making them less costly to transport). Switching to recycled material may be our first thought. After all, we need to find applications for the materials we recover from the waste stream. But reduction of material use is clearly more effective, which may be a difficult truth for advocates of recycling to accept if they have become fixated on recycling for its own sake. Increasing targets for volumes of material to be recycled is a common strategy used by governments around the world. Yet meeting these targets may actually be a result of a greater, not lower, use of materials. The easiest way to recycle more is to use more. If recycling more becomes the end goal, we justify any means of getting there.

The benefits of recycling are not absolute. There can be additional negative impacts arising from recycling itself. It is not always the case that recycling uses less energy than new production. If a product was not designed to be recycled, or at least to allow disassembly at its end-of-life, it may be more trouble than it is worth to seek to do so. A recycling process may also produce new types of waste or pollution. Aluminium is seen as an ideal recyclable material as it retains most of its useful qualities when reprocessed, with a comparatively low energy cost compared to making it new. Air emissions from aluminium recycling can, however, contain toxic metallic chlorides and oxides, acid gases and chlorine gas (UNEP, 2004, p. 29).

THE 3RS: REDUCE, REUSE, RECYCLE

Recycling has its own impacts. We should therefore be ready to weigh up the costs and benefits in any particular case, for it may do more harm than good. The environmental benefits of recycling, and using recycled materials in new product manufacture, are in comparison to not making any efforts to retain those materials or their value, by simply considering them to be waste. Recycling is presented as an alternative to disposing of waste materials in landfill and using virgin materials to make new products. Put that way, recycling does seem a preferable way of dealing with material waste. We instinctively feel that reusing materials from products which would otherwise be discarded must be a good thing, environmentally. Yet disposal is not the only alternative option.

Material recovery and reprocessing is an established industry in developed economies, driven by a combination of environmental ideology, political pressure, legislation and market forces. The UK government published a policy for reducing and managing waste:

> We want to move towards a 'zero waste economy'. This doesn't mean that no waste exists – it's a society where resources are fully valued, financially and environmentally. It means we reduce, reuse and recycle all we can, and throw things away only as a last resort. (DEFRA, 2015)

This statement echoes the Waste Electrical and Electronic Equipment Directive (WEEE), the international product take-back system for the European Union:

> The Directive aims to prevent or reduce the negative environmental effects resulting from the generation and management of WEEE and from resource use. ... [I]ts key purpose is to contribute to sustainable production and consumption by, as a first priority, the prevention of WEEE and, in addition, by the re-use, recycling and other forms of recovery of such

wastes. The Directive thus incorporates the waste hierarchy. (European Commission, 2014, p. 6)

The 3Rs of the waste hierarchy – reduce, reuse, recycle – are the main model for responding to the problem of how to deal with material waste in a society replete with unwanted physical stuff.

Recycling sits third in the familiar three-level waste hierarchy, behind reduce and reuse. It is the least effective of the 3Rs in dealing with the problem of waste (reduce, then reuse, *then* recycle). There is always an energy cost to the reprocessing of materials. This cost is increased by the need to transport materials to suitable and specialist reprocessing facilities, particularly for complex products made from multiple components and sophisticated materials. Aluminium drinks cans and coffee capsules are exemplars of the energy savings possible from recycling used products, as opposed to using virgin material. Manufacturers of these products make claims for almost infinite recycling. Coca-Cola's 'Recyclometer' translates the energy savings into easily understood terms. Recycling one aluminium can saves enough energy to power a lightbulb for 2.5 hours, compared to using new material. This value goes

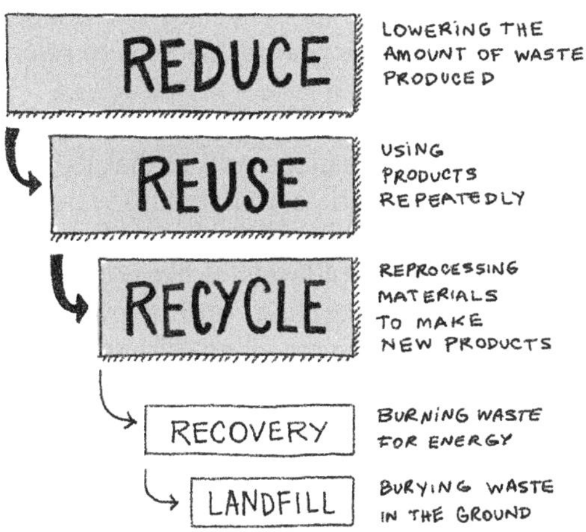

Figure 4.1 The 3Rs and the waste management hierarchy.

down for glass and plastic bottle alternatives. Aluminium is well-known to be repeatedly recyclable with low loss of material quality, but this is in large part due to its high intrinsic material value and high initial embodied energy. Aluminium, the model material for recyclability, is therefore untypical in terms of the benefits that may come from recycling a consumer product. The comparisons made are always between recycling and doing nothing, i.e. straight disposal of items after a single use. The option to not consume is never considered. We might see Coke as the real problem, not simply its cans and bottles. The benefits of a Coke-free world are never considered (Plastic Expert, n.d.).

In the case of glass bottles, greater energy savings could potentially be made by cleaning and reusing (rather than reprocessing) bottles. If this is done locally, transportation impacts are reduced if not removed. This used to be common practice in the UK, with a deposit paid for the return of robust bottles fit for repeat use, and still is in some regions. Elsewhere manufacturers have replaced heavy, reusable glass bottles with lightweight, single-use, non-returnable alternatives in order to reduce their direct financial costs. This is dictated by market forces. Market interventions might therefore direct companies away from such moves, for example through taxation. Initiatives to return single-use bottles for recycling are becoming more common, but these miss the greater benefits of return for reuse (Neslen, 2017).

The most familiar version of the waste hierarchy is the 3Rs: reduce, reuse, recycle. Expanded versions add further Rs such as refurbish, repair, recondition and remanufacture. Each of these additions is a variant on the difficult distinction between reuse and recycling. These additional levels of the waste hierarchy are always inserted above recycling, and below reuse, providing more strategies to prefer to recycling (Figure 4.2).

We might even seek to remove recycling entirely from the waste hierarchy, viewing it as a failure to respect and retain the value of our products (Figure 4.3). Rather than challenging the generation of waste, and reducing the problem, recycling is itself dependent on material excess to keep it going. Recycling, perversely, encourages wastefulness. Because we

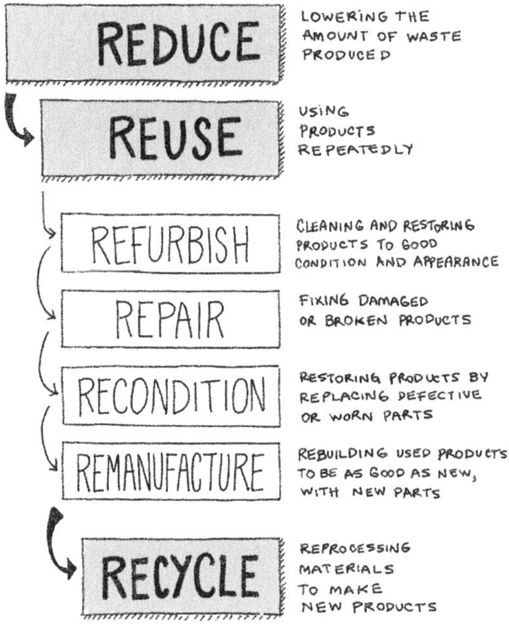

Figure 4.2 The 7Rs of an expanded waste management hierarchy.

believe we can, in principle, recycle almost anything, we appear unconcerned that we are producing too much in the first place. Recycling responds to, but also perpetuates, an attitude of consumption and disposability in relation to material resources. On these grounds we can argue that recycling has no place in a progressive waste strategy. Recycling is omitted from Figure 4.3 because it reinforces, rather than counters, a culture of disposability. 'Refuse' represents the most direct response to wasteful production and consumption, which is the real underlying issue. Challenging the dominant ideology of wastefulness through direct action, such as protest or boycott, is likely to have greater effect than the quiet subversion represented by other approaches further down the list. Replenishing our laundry detergent with the manufacturer's 'eco-refill' pack will slow down very slightly our resource consumption. It does not, however, challenge the principle of centralised production,

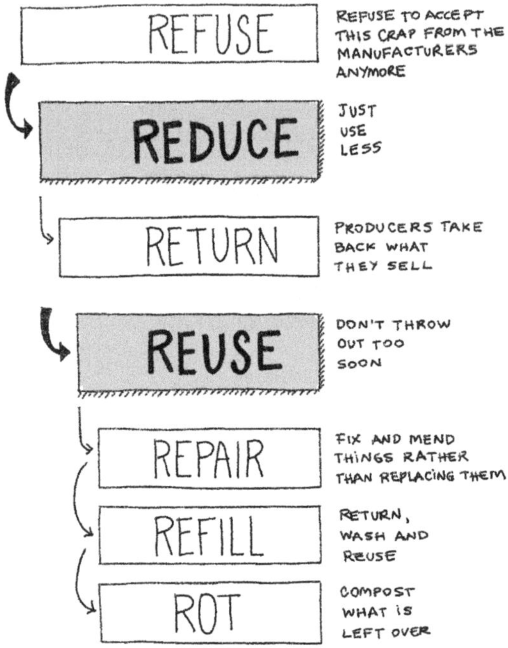

Figure 4.3 Alternative 7Rs (adapted from Alter, 2014).

marketing and distribution of one-size-fits-all commodities which is the underlying driver of our culture of disposability and wastefulness. Despite the fervour with which it is practised by many, recycling cannot really be seen as an activist practice. At best, it represents a minor redirection of a limited volume of the resources in a vast and complex system.

It is only possible to omit recycling from the waste hierarchy, however, if we view things from the perspective of the consumer. From the perspective of the producer, recycling may well follow product takeback at the 'return' stage of the hierarchy given in Figure 4.3. Recycling may be the only option if the product was not designed for anything else. In this case, recycling is not removed from the system, it is just no longer an action that we feel we participate in as consumers.

THE LIMITS OF REUSE

The waste hierarchy, in all its forms, tells us it is better to reuse than to recycle. Reuse sits above recycling, and represents a more sustainable approach to waste management, because it repurposes materials directly and does not require significant additional energy. We should never forget that recycling often has a significant energy cost. To confuse things, what has traditionally been considered as reuse – the repurposing or minimal adaptation of a product – is often now called 'upcycling', as discussed in Chapter 1. Too often upcycling is an essentially reactive and improvised practice which seeks ways to retain the value of waste materials through new and self-consciously creative configurations of unwanted materials. This is a familiar mode of self-styled 'sustainable' design. As we saw in Chapter 2, a UK television show challenged three designers to turn a decommissioned passenger aircraft into novel yet useful new products. The structure was taken apart to be partially remade in myriad new ways. Clocks, lamps, bird boxes, seating and shelters were all made from salvaged pieces of plane (Wollaston, 2014). It is difficult to see this exercise as a useful response to the problem of waste, however. It does not seek to address the production of waste in the first place, through the elimination or avoidance of industrial leftovers. Nor does it encourage design for reuse, reconditioning, refurbishment, repair, remanufacture or indeed further recycling, as suggested in Figures 4.1 to 4.3. Waste is, after all, a design flaw, and designing-out the production of waste in the first place (i.e. 'reduce') is the best response. Uncritical celebration of upcycling may actually do harm if it distracts us from more meaningful ways of addressing material waste. Upcycling, as a mode of reuse, does not address the systemic causes of waste production, only its most visible effects.

As discussed in Chapter 2, the Swiss company FREITAG repurposes truck tarpaulin to make bags and other accessories. FREITAG aligns itself with sustainability:

> It's all about cradle-to-cradle. We can also talk about recycling, upcycling and downcycling. There are organic cycles, there are technical cycles. And

> we think it's important to consider these cycles, especially before you start to mix up these materials. So you pay attention and say, 'ok, what are the materials I use?' Is it a material that I can bring in the compost or throw in the backyard and something new grows out of it? Or do I choose a technical cycle? So thinking about those cycles is one of the environmental aspects, and I think the world needs solutions like this. (Seeds and Fruit, 2009)

Daniel Freitag here invokes the influential Cradle-to-Cradle design protocol, and its distinction between technical nutrient cycles and biological (here 'organic') nutrient cycles (McDonough & Braungart, 2002). FREITAG's approach to material reuse is, however, entirely reactive to existing waste streams, over which the company has no real control. If tarpaulin were replaced as the material of truck covers, where would that leave FREITAG? Cradle-to-Cradle means designing cyclical and integrated manufacturing systems in their entirety, not simply responding to the waste streams of current linear manufacturing systems, of which material waste is an accepted but unwanted by-product. FREITAG's products have clear symbolic value – the company has a strong brand identity comprised in part of its creative reuse of waste industrial product. Yet its quantifiable impact in diverting waste from disposal in landfill is small in relation to the size of the waste stream. Visible examples of material reuse may of course stimulate interest in material cycles and waste. But examples like FREITAG beg the question: should we really be making bags and wallets from recovered truck tarpaulins? FREITAG and others are not making a significant contribution to the redesign of the industrial production and consumption system which is actually called for by Cradle-to-Cradle. A FREITAG bag may actually be a distraction from the real sustainability project, in the same way that carbon offsetting distracts us from actually addressing the direct environmental impacts of aviation. When we pay a premium to offset the carbon emissions arising from our occupying a seat on a commercial flight, we do not actually reduce those emissions, or even directly address the harm done by them. In positioning themselves as suppliers of consumer lifestyle products, FREITAG are as complicit in generating consumption as any other manufacturer. Many recycled products simply postpone the point at which their materials are disposed of in the conventional way – landfill – rather than contributing to the perpetuating material cycles advocated by Cradle-to-Cradle thinking. Daniel Freitag is right, the world does need solutions of the kind he discusses in the interview above, but messenger bags made from recovered truck tarpaulin are not one of them.

If a FREITAG bag does not represent a serious attempt to allay waste arisings, we might view it instead as an example of 'critical' design:

> Critical Design uses speculative design proposals to challenge narrow assumptions, preconceptions and givens about the role products play in everyday life. (Dunne & Raby, 2014)

Design is used as a critical practice, to create design fictions which are provocative yet plausible. A premium-priced consumer lifestyle product made from salvaged truck tarpaulin and seat belt, promoted using aspirational rhetoric, seems to fit the bill as an example of this mode of design-as-provocation. As an accidental critical design, a FREITAG messenger bag ridicules our notions of consumerist products, whilst also being available to buy. On this reading, the entire FREITAG project – encompassing the products, the company and the brand story – might be viewed as an exercise in revealing the wrong-headedness of seeking to address high-volume waste streams by dipping into them to make comparatively very low volumes of high-end consumer products. With roots in critical social theory, critical design is always done self-consciously and deliberately. Much design with recycled materials is, in contrast, well-intended but naive about the deeper implications of the collective recycling project to which it contributes. Much recycled product design is faux-sustainability. Yet in highlighting unsustainability, or environmental tokenism, such products are nonetheless useful in showing us an inadequate response to the challenge of sustainable waste management. Observers, however, typically lack the sustainability literacy to recognise them as such, and champion FREITAG and others as exemplars of sustainable design practice. The use of recycled or reclaimed materials is a common entry point into sustainability-led thinking and practice for designers, who are encouraged to think of themselves as the definers and makers of our material culture. A designer's first self-consciously 'eco' project is often a familiar product remade with a reclaimed or recycled material. If that is where her engagement with sustainability ends as a creator of our material culture, then little progress will be made in addressing the deeper systemic unsustainability of which waste streams are the most visible symptom. Sustainability is a systems-based concept which requires us to take a holistic view of issues which we may be used to considering in much more simplistic terms. As a maker, it is natural for a designer to approach sustainability via the materials she uses, and using waste materials (especially those which have already been through one or more phases of industrial production and consumption) feels intuitively right. Yet by perpetuating an inherently wasteful materials economy, the designer using recyclate may become an unwitting accomplice of the system she seeks to question.

THE UNSUSTAINABILITY OF RECYCLING

Unsustainability has deep roots in our societal system, and limited change can come from individual action.

> While recycling is by far the most common practical step that people take to help the environment ... other problems, much more difficult to address at the household level, are clearly more urgent than recovery of materials from trash. (Ackerman, 1997, p. 2)

Recycling is entry-level environmentalism, but it seems achievable. It is also widely taken as an indicator of broader environmental awareness and pro-environmental action. To recycle is to be seen to be green. Recycling will not in itself save the Earth, or our own chances of survival upon it. Yet many of us behave as if we believe that recycling is enough, that if we recycle we need not do very much more in order to move towards a lifestyle which satisfies the requirements of sustainability. This view is problematic. If we consider environmental impacts throughout the whole process of production and consumption, and look beyond the simple preservation of materials, recycling can actually do more environmental harm than good. Recycling can be unsustainable – as when, for example, collected material is transported around the world to be reprocessed. At worst, an exclusive focus on recycling distracts us from more effective courses of action, and represents a simplistic response to the complexities of sustainability.

Developed consumer economies pay other parts of the world to take care of their recycling. Waste is traded in a global recycling industry, the movement of unwanted products and materials matched by a movement in money. The processing of electronic waste at its end-of-life is carried out far away from the consumer markets in which the products were used. Computers are torn apart by hand, their most valuable raw materials recovered by the crudest methods, with severe costs to human health. People working at this informal recycling are exposed to terrible hazards, reclaiming residual value from products which were never intended for recycling. Electrical and electronic waste streams are growing rapidly worldwide, meaning more people are forced to work in awful conditions on our behalf elsewhere in the world due to our consumption (World Economic Forum, 2019). If it is economically viable for waste materials to be recovered, they will be, somewhere in the world. As with the exploitative production of cheap clothing, human costs are acknowledged but generally overlooked in the global recycling industry. Yet sustainability in its broad sense is concerned with social and environmental considerations. We should be offended not just by the environmental pollution arising from informal recycling of e-waste, but by the fact that this is the workplace of people with no other option to earn a living. As discussed in Chapter 2, the aesthetics of recycling derive from the entire system of actions, affordances and behaviours by which materials are recovered for reuse. Globally, this is a system in which physical materials are granted higher value than human wellbeing. The violence done to e-waste recyclers by the perils of their labour is direct and immediate. The contrast between the sophistication of a computer in use and the crudeness of its treatment at end-of-life is stark. Once it reaches a burning field in Ghana, what was recently a valuable consumer product is give the most brutal treatment to recover one of its basic component elements,

Figure 4.4 E-waste.

copper (Goutier, 2014). These harsh realities are an undeniable part of the contemporary global recycling system, and difficult to square with the notion that recycling serves an agenda of social and environmental sustainability.

Recycling is a global industry in which waste products are a commodity to be traded just like any other. As with other industries, recycling is driven by economic logic. If it makes financial sense to transport waste around the world for reprocessing, this will happen. The environmental and social implications of such financially viable practices are rarely considered. Even when we could recycle closer to home, we outsource it to those who can do it cheaper. Abiding by a principle of sustainability, we would process recyclables locally to their place of consumption. Instead, they are shipped to nations with lower labour costs and safety standards, to be remade into new goods and returned to markets in Europe and North America. Nations in Asia and Africa are the greatest industrial recyclers in the world, but not the greatest consumers (Leach & Boyd, 2017; *The Economist*, 2007). Shipping waste around the world in huge quantities makes little sense environmentally due to the transportation impacts. We are told to recycle for the sake of the environment. Globally distributed recycling diminishes the value of our recycling efforts. Careful sorting and collection of our waste is a personal and collective civic duty. The value of our actions is, however, reduced with every kilometre our waste is then transported to be dealt with. The inherent unsustainability of global circulation of waste in the name of recycling compels us to consider a more local response. Sending our waste to the other side of the world lets us off the hook. Dealing with it more directly would lead us to greater self-examination as to why and how we generate so much waste.

Recycling may not always be the 'practical step to help the environment' we think it is (Ackerman, 1997, p. 2). Recycling is convenient for producers and consumers alike, as it fits into our existing practices. Through recycling, production of waste from the factory to the household can be managed as an additional resource flow. Recycling is privileged over reduction and reuse because it is less disruptive. Waste management companies and their waste collection trucks promote recycling as the default approach, with little mention of options above it in the waste hierarchy. Reuse is supplementary to recycling, more difficult to achieve despite being a preferable option. The renaming of reuse as upcycling, discussed above, indicates its demotion to a specialist sub-category of recycling.

We feel instinctively that recycling is worthwhile. Recycling also demands little of us. So we put our faith in the principle of recycling, even if we are not always confident of the value of the practice. We trust that recycling is beneficial much of the time, and this overrides reservations about its value in a particular instance – where does this multi-material plastic bottle actually go from here? Yet even if recycling is broadly environmentally beneficial, it still occupies too much of our environmental awareness. If all we do is recycle, we are engaging in environmental tokenism, as in when we diligently drop an empty water bottle into a recycling bin at the airport before boarding a long-haul flight. At its worst, a blind commitment to recycling fuels a naivety which bodes ill for our ability to move collectively to a more sustainable future. The fact that almost anything can, in principle, be recycled does not necessarily mean that it should be. A belief that everything is recyclable enables a continuation of the disposability which is central to our consumer culture. Our practices of recycling and reuse do not currently make a significant quantifiable contribution to addressing waste generation and disposal. The present scale of our attempts at recycling and reuse is small in relation to the volumes of waste generated by our lifestyles. Significant change can only come from a systemic response to waste, and a wider recognition that the most-preferred strategy for dealing with the problem of waste is the first of the 3Rs – reduce.

The institutionalisation of recycling as an industrial and cultural practice may actually prevent us from meaningfully engaging with sustainability, if we think that everything will be taken care of by the system. Recycling may therefore drive unsustainability. This is the paradox in the development of recycling as an industry. A thriving recycling system requires stable, even increasing, volumes of waste as its input. The processes of recycling need feeding, like any other industrial process. In this case the feedstock is material waste, the output of a linear system of production and consumption. A successful recycling system therefore not only responds to waste streams: it actually encourages the generation of further waste to perpetuate itself within the growth logic of capitalist economics.

The opening scenes of the 2008 Disney Pixar film *WALL-E* show the fastidious waste collection and sorting efforts of the title character. WALL-E (Waste Allocation Load Lifter – Earth-Class) is a battered trash-compactor robot, programmed to unthinkingly collect and sort the material waste left behind by departed humans:

> It is 700 years in the future. A city of skyscrapers rises up from the land. A closer view reveals that the skyscrapers are all constructed out of garbage, neatly compacted into squares or bales and piled on top of one another. In all the land, only one creature stirs. This is WALL-E, the last of the functioning solar-powered robots. He – the story leaves no doubt

> about gender – scoops up trash, shovels it into his belly, compresses it into a square and climbs on his tractor treads and heads up a winding road to the top of his latest skyscraper, to place it neatly on the pile. (Ebert, 2008)

Humour arises from WALL-E's confused interactions with various objects and relics of the human consumer culture of the film's past (our present). The robot does not know the intended purpose or value of the objects he collects, but gives them meaning and use in his world. WALL-E considers himself to be engaged in meaningful and constructive work – collecting and sorting the detritus of a decaying human-made world. The question of what is to be done with the material he so diligently collects and sorts is not addressed, however. Without any clear purpose beyond a quest for tidiness, WALL-E's commitment to his task appears to be a form of Obsessive Compulsive Disorder, manifest as extreme hoarding behaviour. The purposelessness of WALL-E's efforts can be seen as a comment on the dubious value of much of our own recycling activity. The film illustrates the distinction between initial material recovery and collection, and subsequent recycling and reuse. Recovering materials is a form of stockpiling. Unless we then do something useful with those materials, which ideally preserves their material value without any further ecological costs, we are not acting in the interests of sustainability. WALL-E's quest to tidy up his world for personal and aesthetic reasons therefore potentially mirrors our own. How can we ensure that our own commitment to recycling is not judged similarly? We must honestly critique and evaluate our recycling, and be prepared to see the limits of its usefulness. We must acknowledge when it acts against the interests of sustainability.

Recycling has power as a metaphor, which the film *WALL-E* makes great use of. It symbolises circularity, and the notion that we should not take, use and dispose of materials in linear sequence. This metaphorical value does not justify our emphasis on recycling as the primary strategy for dealing with the problem of waste. Recycling targets relate only to volumes of waste material recovered from waste streams, not what we then do with it. Much recycling activity is driven by a simplistic focus on conserving materials, ignoring other environmental and social impacts of material reprocessing and reuse, and fails to take a holistic and systems-based view of the sustainability (or unsustainability) of recycling practices.

Recycling does not feature in the top ten actions we are recommended to take to reduce our carbon footprint (Harrabin, 2020). Recycling typically reduces emissions by 0.01 tonnes of CO_2 equivalent per person per year. This contrasts with much higher reductions for other lifestyle changes, including 1.68 tonnes from taking one less long-haul flight per year, and 1.6 tonnes from switching to renewable household energy supply. To put these figures in context, current annual household emissions are 10 tonnes in the UK, and 17 tonnes in the US. It is notable that an accompanying chart showing figures for the top ten options represents renewable energy supply using a recycling symbol, even though recycling itself is proven to be a minor potential contributor to a more sustainable lifestyle (Harrabin, 2020). As in the case of the UK bank discussed in Chapter 3, the recycling symbol continues to do much more work than it really should, to the detriment of a critical understanding of its value.

DESIGNING-OUT WASTEFUL BEHAVIOURS

Designing-out waste is the first choice in the waste hierarchy (Figures 4.1 to 4.3). If waste is a design flaw, design can help us address it. This is more difficult than recycling or reuse due to the scale and complexity of the systems by which we produce and consume things. The decision to redesign a product means redesigning all its processes of production, and is not the designer's alone. There are many factors to consider – financial, social, political, legal. It is therefore difficult, costly and risky to redesign any product. For a manufacturer, it is easier to try to deal with waste further down the line (the 'end-of-pipe' approach) than eliminate it upstream in the production process. It is easier to design a recycling programme than redesign a product (Ackerman, 1997, p. 6). Yet to do so is to view waste as an acceptable by-product, as an unwanted output to be managed rather than eliminated. This is fine if we have effective and sustainable ways of dealing with waste at the end of the production–consumption process, but we generally do not. Reuse and recycling are both unsatisfactory in dealing with the problem of waste. Rather than addressing the production of waste, they seek to soften its impact, and in effect excuse its production. Recycled product manufacture at any significant scale requires steady and reliable supplies of recyclate. This requires a waste collection and reprocessing industry which will only be viable if there is an assured long-term generation of waste. Recycling at scale encourages, rather than reduces, the production of waste. It itself becomes just another mode of production, with its own associated environmental and social impacts.

This encouragement of waste generation is the rebound effect of recycling, and influences our individual consumption behaviours. If we can easily recycle, we use more paper products than when recycling is not available (Catlin & Wang, 2013). Such perverse outcomes are not intended by the designers of recycling programmes, but must be acknowledged. Accessible recycling systems can be counter-productive by excusing us from considering the costs of our wastefulness. If paper towels in a public toilet can be readily collected for recycling via a provided bin, why not use as many as you like? When we participate in recycling by correctly disposing of an item, we feel good about ourselves in observing a positive social norm. If we are constantly told to recycle, we feel virtuous when we do. Taking this to its logical extreme, we can generate more recycling opportunities by deliberate, or at least thoughtless, over-consumption. At that point recycling becomes entirely disconnected from reasons for doing it. It becomes a virtuous act which has value in and of itself, irrespective of whether or not we really needed to use the item that we are now recycling. This may be compounded by a 'licensing effect' whereby having recycled something we feel we can indulge in non-environmental behaviour in return, such as car driving. A supposedly virtuous act (recycling) absolves us of guilt for a clearly non-virtuous act (driving). This is the opposite of the hoped-for 'spillover' effect whereby one pro-environmental behaviour (recycling) leads to more in other areas of our lives (cycling rather than driving) (Thøgersen & Crompton, 2009).

When the pro-environmental action is relatively inconsequential in comparison to the non-environmental response, careful but tokenistic recycling has a negative effect on our overall pro-environmental behaviour. Living car-free is in first place in the list of actions we are recommended to take to reduce our carbon footprint, reducing emissions by 2.04 tonnes of CO_2 equivalent per person per year; the figure for recycling is 0.01 tonnes (Harrabin, 2020). The figure for switching to a battery electric car is 1.95 tonnes, but as this considers emissions and not the other impacts of this flawed alternative technology, non-fossil-fuel driving is similarly hard to justify (The Conversation, 2018).

The relation of recycling to consumption is complex. In developed economies, recycling is most strongly associated with the affluent middle classes and older people, two groups which overlap (Martin *et al.*, 2006; Langley, 2012). Affluence is expressed through conspicuous consumption, for which we then seek penance by recycling. Those who recycle most have most to recycle. Thrifty consumption, for example buying only what we know we need and will use, will generate less waste than indulgent consumerism, yet the latter provides greater opportunity for recycling. In a culture where to be seen to recycle is good, giving ourselves more opportunity to do so by over-consuming has a perverse justification. Recycling rates are often taken as a proxy for progress on sustainability, yet may be nothing of the sort; 'municipal recycling participation is by no means a good proxy for other forms of [pro-environmental] practice' (Barr *et al.*, 2013, p. 72). Rising recycling rates tell us nothing without reference to baseline waste arisings, and the effectiveness of the recycling processes used. Headline recycling rates distract us from engaging more deeply with the challenges of collective sustainability. These statistics capture formal or municipal recycling activity; informal and domestic reuse activities are invisible to those who gather the data. Keen participation in recycling can be negative if it is used to license other environmentally detrimental actions. The environmental gains from recycling can be small, and the accompanying losses great.

In developing economies, rising material waste streams indicate an escalation of consumer culture and the rise of a middle class. Material consumption increases in proportion to affluence. Around the world, an increase in financial wealth is usually accompanied by an increase in material waste from consumption of material goods. Given this, is sustainability compatible with affluence (Ackerman, 1997, p. 6)? Not if recycling is the answer. The central principle of recycling is preservation of materials and their intrinsic value, beyond the lifespan of their embodiment in a particular product. There is tension between this and the consumerist principle that we can use and discard products and materials however we like, if we are rich enough not to miss them. Recycling seeks to resolve this tension by allowing us to consume to our hearts' content, hopeful that we can return the things we no longer value to a collective materials pool that feeds new cycles of production.

CONCLUSION

'Waste not, want not' is a principle from another time, with unwelcome associations of poverty and deprivation. We now have recycling to excuse our wastefulness. Yet this is likely to have limited impact within an unchanged consumerist system. Recycling within an ecomodernist paradigm can bring no more than eco-efficiency. It provides a means of fine-tuning an inherently unsustainable system by making it less bad. 'Sustainable' is commonly used as a prefix to try and shift our perception of almost anything. Yet some products are so inherently unsustainable in the form we know them that this is nonsense. There cannot be sustainable aviation, using fossil fuel-based technologies. There cannot be sustainable fishing, on an industrial scale. There cannot be sustainable products, made in the same way as conventional products but from recycled materials. Recycling as we know it aids and abets a system which is irreparably flawed in terms of prospects for improved future sustainability. Production and consumption must be considered together when we reflect on how to most effectively address the issue of waste. Recycling occupies third-best position in the most popular version of the waste hierarchy, behind reduction and reuse. A commitment to recycling for its own sake can generate more waste, as the act of recycling becomes detached from a rationale for doing it. If recycling can be a driver of unsustainability, it is then possible to be a non-recycling environmentalist. The next chapter considers the future prospects for recycling as a worthwhile waste strategy, focusing specifically on design for recycling.

REFERENCES

Ackerman, F., 1997. *Why do we recycle? Markets, values, and public policy*. Washington, DC: Island Press.

Alter, L., 2014. Today is America Recycles Day, the annual greenwashing homage to a culture of disposability [online]. Available from: https://www.treehugger.com/i-almost-missed-america-recycles-day-4851629. [Accessed 1 September 2020].

Barr, S., *et al.*, 2013. Beyond recycling: An integrated approach for understanding municipal waste management. *Applied Geography*, 39, 67–77. https://doi.org/10.1016/j.apgeog.2012.11.006. [Accessed 1 September 2020].

Catlin, J., and Wang, Y., 2013. Recycling gone bad: When the option to recycle increases resource consumption. *Journal of Consumer Psychology*, 23(1), 122–127.

The Conversation, 2018. Not so fast: why the electric vehicle revolution will bring problems of its own [online]. Available from: The Conversation, https://theconversation.com/not-so-fast-why-the-electric-vehicle-revolution-will-bring-problems-of-its-own-94980. [Accessed 16 January 2021].

Cooper, T., 1994. *Beyond recycling: The longer life option*. London: New Economics Foundation.

Department for Environment, Food and Rural Affairs (DEFRA), 2015. *2010 to 2015 government policy: Waste and recycling* [online]. Available from: https://www.gov.uk/government/publications/2010-to-2015-government-policy-waste-and-recycling/2010-to-2015-government-policy-waste-and-recycling. [Accessed 1 September 2020].

Dunne, A., and Raby, F., 2014. Critical Design FAQ [online]. Available from: http://dunneandraby.co.uk/content/bydandr/13/0. [Accessed 1 September 2020].

Ebert, R., 2008. Droid story [online]. Available from: https://www.rogerebert.com/reviews/wall-e-2008. [Accessed 1 September 2020].

The Economist, 2007. The truth about recycling. *The Economist*, 383 (8532).

European Commission, 2014. Frequently Asked Questions on Directive 2012/19/EU on Waste Electrical and Electronic Equipment (WEEE) [online]. Available from: http://ec.europa.eu/environment. [Accessed 1 September 2020].

Goutier, N., 2014. E-waste in Ghana: Where death is the price of living another day [online]. Available from: https://theecologist.org/2014/aug/07/e-waste-ghana-where-death-price-living-another-day. [Accessed 1 September 2020].

Harrabin, R., 2020. Climate change: Top 10 tips to reduce carbon footprint revealed [online]. Available from: https://www.bbc.co.uk/news/science-environment-52719662. [Accessed 1 September 2020].

Langley, J., 2012. Is green a grey area? Sustainability and inclusivity: Recycling and the ageing population. *The Design Journal*, 15(1), 33–56.

Leach, A., and Boyd, O., 2017. Samsung and Greenpeace: What you need to know about e-waste [online]. Available from: https://www.theguardian.com/sustainable-business/2017/mar/01/samsung-greenpeace-what-you-need-to-know-e-waste-smartphones-recycling. [Accessed 1 September 2020].

Martin, M., Williams, I. and Clark, M., 2006. Social, cultural and structural influences on household waste recycling: A case study. *Resources, Conservation and Recycling*, 48, 357–395.

McDonough, W., and Braungart, M., 2002. *Cradle to Cradle: Remaking the way we make things*. New York: North Point Press.

Neslen, A., 2017. Coca-Cola U-turn could help UK catch up on can and bottle recycling [online]. Available from: https://www.theguardian.com/sustainable-business/2017/feb/28/coca-cola-u-turn-can-and-bottle-recycling-europe. [Accessed 1 September 2020].

OpenIDEO, n.d. How might we establish better recycling habits at home? [online] Available from: https://challenges.openideo.com/challenge/recycle-challenge/brief. [Accessed 1 September 2020].

Plastic Expert, n.d. Plastic Expert reviews the Coca Cola Recyclometer [online]. Available from: https://www.plasticexpert.co.uk/recycling-reviews-coca-cola-recyclometer/. [Accessed 1 September 2020].

Seeds and Fruit, 2009. FREITAG – customized bags [online]. Available from: http://www.seedsandfruit.com/2009/10/freitag-customized-bags. [Accessed 1 September 2020].

Thøgersen, J., and Crompton, T., 2009. Simple and painless? The limitations of spillover in environmental campaigning. *Journal of Consumer Policy*, 32, 141–163.

United Nations Environment Programme (UNEP), 2004. Vital waste graphics [online]. Available from: http://www.grida.no/publications/vg/waste/. [Accessed 1 September 2020].

United States Environmental Protection Agency (US EPA), 2020. *Basic information about landfill Gas* [online]. Available from: https://www.epa.gov/lmop/basic-information-about-landfill-gas. [Accessed 1 September 2020].

Wollaston, S., 2014. Kevin's Supersized Salvage – TV review. *The Guardian* [online] https://www.theguardian.com/tv-and-radio/2014/apr/25/kevins-supersized-salvage-tv-review. [Accessed 1 September 2020].

World Economic Forum, 2019. *A new circular vision for electronics: Time for a global reboot.* Geneva: WEF.

Young, J., and Sachs, A., 1994. *The next efficiency revolution: Creating a sustainable materials economy.* Worldwatch Paper 121. Washington, DC: The Worldwatch Institute.

5

FUTURES

REACTIVE RECYCLING

Designing with recycled materials is often reactive. We are constrained by decisions made in the design of an original product for which we now seek a new use. Recycled design is also often dependent on unreliable or temporary waste streams. A particular consumer product can provide a brief opportunity for creative material reuse before it is discontinued. Reactive recycled design is always vulnerable to unforeseen changes made by the manufacturers of source products. Suppliers of consumer products increasingly have a 'takeback' policy, by which they will collect your old product when delivering its replacement, even if it is not one of theirs. Yet, as has been emphasised throughout this book, collection of unwanted or waste materials does not in itself constitute recycling.

The Reee chair was launched in the UK in 2008 by Pli Design. It was an early example of an industrially designed and manufactured recycled content product with clearly communicated environmental lifecycle savings over an equivalent product made from non-recycled materials. The Reee is named after the European Waste Electrical and Electronic Equipment (WEEE) Directive, legislation which gives manufacturers a responsibility to take back the products they make once they are discarded by consumers. The Reee is a product design responding to a specific waste stream – the Sony PlayStation 2 (PS2) computer games console. The claims made by the manufacturer of the chair are impressive. All plastic used in the Reee is 100% recycled post-consumer waste. The plastic recycling is efficient due to the low energy input required. This recycled plastic is itself 100% recyclable, due to the high-quality single-source material used. The chair uses modular assembly, making it easy to repair and refurbish by replacing easy-to-remove elements as needed throughout the chair's lifecycle, enabled by Pli's own takeback scheme for the Reee. The chair's powder-coated steel frame has a relatively low environmental impact in its manufacture. The material economy of the Reee is also local – games consoles are collected, their material is reprocessed and the re-formed plastic elements are used in new product manufacture entirely within the UK. Transportation is often a significant contributor to a product's environmental impact; many things are made far away from where they are sold. As the Reee is sourced and manufactured entirely in the UK, it travels a fraction of the distance from factory to customer of a chair made in Asia and shipped to Britain, for example (Pli Design, 2008). We should of course remember that the original games consoles were not made locally, and are themselves a product of a globally distributed production and consumption system. We do not generally hold recycled products accountable for the impacts of the products from which they are sourced. We wipe the slate clean when we compare them to brand-new products made from newly extracted materials. A recycled product, after all, in theory causes a brand-new product to not be made, or at least not consumed. By this usual reckoning, the Reee chair is a model of ecodesign, combining a holistic approach to material recycling with low-impact manufacture and local sourcing.

The aesthetics of the Reee chair are different to many other products of recycling – its material origins are not visibly apparent. We would not know the material story unless told about it. 'Recycle again … and again … and again …' is discreetly imprinted on the back of the chair, but this is only apparent to the particularly curious. Compare this with a FREITAG bag, the ultimate messenger for recycling, whose reused materials are an explicit visual element of the remade product.

The Reee chair helps Sony meet its obligations for product recovery and reuse, via collaboration with a small specialist furniture manufacturer. Cooperation between these two quite

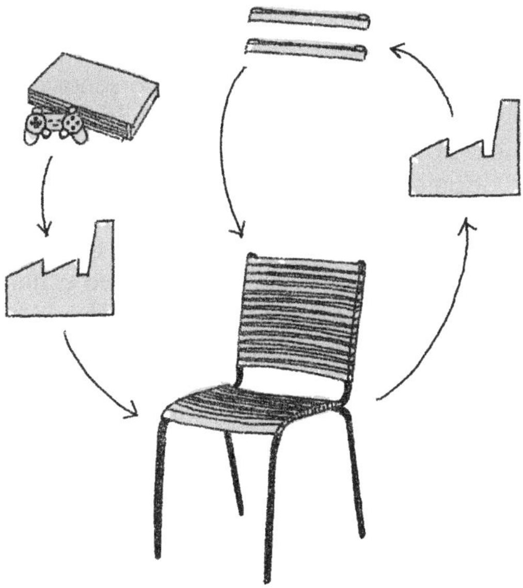

Figure 5.1 Reee chair.

different companies helps to deliver the intentions of the WEEE Directive, which seeks to enforce extended producer responsibility within Europe. This example shows the potential for legislation to prompt a change in industry practice which benefits the companies involved. Sony meets some of its legal obligation, and receives some positive press. Pli gains new business and heightened profile as a small design and manufacturing company. The Reee is, however, clearly limited in its potential impact. Pli Design consciously established a local model for material recovery and new product manufacture. It would not make sense to expand the reach of this UK model; better to export the model elsewhere so that local waste is used for local production in other places. More fundamentally, the Reee is vulnerable due to the specificity of the waste stream to which it responds. The PlayStation 2 video game console was released in Japan in March 2000, becoming the world's best-seller with over 150 million units sold worldwide. PS2 continued to be produced until January 2013, but its successor,

PlayStation 3, was introduced in 2006, marking the demise of the console used for the Reee chair (Stuart, 2013). According to the founder of Pli Design, Christopher Pett (personal communication, 8 July 2015), approximately 1,000 Reee chairs were made between 2008 and 2011, reusing the plastic from around 8,500 games consoles. The Reee is no longer in production because the supply of recyclate, in the form of PS2 console casings, ran out. PS2 was a long-lived consumer electronics product, yet was always destined to have a finite life, to be inevitably replaced by a newer version. This is the challenge for a design approach which reacts to a particular waste stream. Increasingly rapid innovation in consumer products means that any secondary product created in response will have a limited production lifespan. Technologies used in home audio and televisions have become unpopular and then obsolete in short timespans. As consumers we make collective transitions from old to new technologies which generate large but temporary waste streams, as most of us get rid of our old TVs at more or less the same time. The opportunistic designer then has a brief window in which to exploit a glut of discarded products and materials, before supply dries up. Once everyone has replaced their last-generation TV, the supply of dead products will stop. The recycling designer therefore needs to have an eye on the next upcoming waste stream. Most new products, particularly those which are technology-based, will soon be obsolete, becoming the next source of materials for recovery and recycling.

Production of the Reee chair required a stable and continuing supply of waste PS2 consoles. The modular design and assembly of the Reee allow individual elements in the seat and back to be easily replaced. As these elements are themselves recyclable, it is theoretically possible to maintain the existing population of chairs using only the material currently in use. Fresh supply of replacement material from newly sourced consoles would, however, be needed for ongoing production of new chairs. A relationship between Sony and Pli Design produced a useful response to a particular and visible consumer waste stream. Whilst the Reee demonstrates an encouraging principle, it does not, however, tackle the magnitude of the waste stream; a tiny fraction of the millions of PS2 consoles sold were used in Reee chairs.

The Brighton Waste House, completed in 2014, is an example of reactive design with waste materials to make something bigger than a chair (BBM Sustainable Design, n.d.). The main structure of the building is rammed earth chalk wall made from 10 tonnes of chalk waste, with 10% clay. Ramming earth together to make walls is a traditional building method with high thermal insulation. The wall cavities of the Waste House are filled with household and consumer waste products including plastic razors, denim jeans, DVDs, video cassettes and toothbrushes used once by business-class and first-class airline passengers. The building is completed by bricks, plywood sheets and off-cut timbers salvaged from local construction projects. The Brighton Waste House is therefore largely made from familiar household and construction waste. The technical performance of the building in use is monitored, to establish the functional value of its use of waste materials. The project of course also has symbolic value, as an attempt to design and construct a two-storey building from local waste. We should consider carefully, however, what the Waste House communicates. Using surplus consumer products as infill for walls seems crude when we consider how carefully they were conceived and created. Downcycling products with high material value and embodied energy is relatively easy to do, and even easier to communicate to a general audience. If it condones the single use and disposal of a highly durable and industrially manufactured product like a plastic toothbrush, however, the Waste House does little to improve our relationship with resources. Repurposing a product for its material properties is better than throwing it away, but we should acknowledge the wasteful practices which supply an output like the Waste House. If we now start to build using discarded consumer products, we need to continue consuming and discarding those products to sustain that practice. If our use of waste increases, our creation of waste must keep pace. As we recycle more, we must have more to recycle – this is the self-perpetuating internal logic of a growth model of recycling. Putting consumer products into wall cavities clearly fails to make the most of their potential value. It is easier, however, than designing-out waste by rethinking our current linear approach to resource use.

Reactive design with waste materials, in chairs or buildings, offers limited scope for more sustainable production and consumption. Greater impact can come from a more strategic integration of original and secondary production, in which a response to future waste streams is designed-in to initial manufacture.

STRATEGIC RECYCLING: CRADLE-TO-CRADLE

A product is a temporary configuration of materials in a form that delivers some current benefit. A product's useful life may end for functional, technological or stylistic reasons. Effective capture of materials at end-of-life requires forward-thinking design strategies such as modularity and design for disassembly. A product's component materials can thereby be made easily available at the product's demise, to be reconfigured in new and perhaps unforeseen products. We may not always be able to future-proof a product, as we cannot always predict where technology and innovation might lead, but we can seek to future-proof the product's ability to support new uses of its materials.

Herman Miller is a leading international manufacturer of office furniture, equipment and home furnishings. The company has an environmental design protocol which guides its product design across four areas: (1) Materials chemistry – are the chemicals used the safest available? (2) Disassembly – can products be taken apart at the end of their useful life to recover their materials? (3) Recyclability – in addition to having recycled content, can materials be recycled at the end of a product's useful life? (4) Life Cycle Analysis – is the entire lifecycle of a product considered? (Herman Miller, 2020). The Embody performance work chair is a product of these design priorities. According to Herman Miller, this chair has 44% recycled content (mostly post-consumer), contains no PVCs (a widely used but environmentally harmful synthetic plastic polymer) and is 95% recyclable (depending on local facilities). These overall percentages of recycled and recyclable content are aggregates of variable figures for the chair's different component materials. The steel has 37% recycled content and is 100% recyclable, the aluminium is 100% recyclable, and several textiles are available with 100% recycled content (Herman Miller, 2014). It is notable that Herman Miller puts more emphasis on the recyclability of its materials than the recycled content of its products. Contributing to perpetual material cycles is considered more valuable than finding applications for existing secondary materials. Embracing recyclability means strategically designing *for* recycling, so that the material ingredients of a product can be diverted away from waste streams and towards recycling systems. The new material economies created may be 'closed loop' – recovered materials are incorporated into new production of the same products, or 'open loop' – these materials feed into wider material streams shared with other manufacturers. Closed loops are managed separately by individual companies. Open loops require coordination between companies, to share materials effectively. Both require systems-level thinking to facilitate effective recovery and reuse of materials across time and space, involving multiple stakeholders. Herman Miller's foam and textile materials are part of an open loop system which allows them to be recycled into everything from automotive components to carpet padding at the end of their first useful life in a chair (Herman Miller, 2014, p. 2).

DESSO, a leading global supplier of carpeted flooring, uses a closed loop model. This has two requirements: (1) being able to take back used goods and then recycle or reuse the materials; (2) making products with the purity to be recycled at high levels (Tarkett, 2020). DESSO remains the perpetual owner of the materials in its products. The materials are on loan to customers for as long as they use the product. DESSO reclaims ownership of its materials when the product is returned to the company at the end of its service period. This model is enabled via a leasing arrangement, rather than customer purchase of carpets. It is easier to take back a product when it is only on loan to its user. DESSO emphasises the importance of material quality for effective recycling. Waste streams which preserve material purity and integrity, without contamination by inferior

Figure 5.2 Herman Miller's Embody chair.

or alien materials, are key for successful recycling (rather than downcycling). One partially full coffee cup can prevent several tonnes of waste paper being recycled due to contamination and loss of material quality. This is even more important with high-grade materials. DESSO's takeback scheme allows it to control what returns from the customer, to be fed into its own new products via a closed loop of material recycling.

These examples of strategic recycling go beyond one-time reuse of recovered content. They aspire to ongoing 'multicycling', achieved through design for recycling and inherent material recyclability (Duvaltext, n.d.). Designing for perpetual material cycles is codified in the Cradle-to-Cradle design protocol, which sees product design as a form of materials chemistry (McDonough & Braungart, 2002). In this framework, products are designed to be easily dissolved and reconstituted at the end of their useful life, with minimal loss of material value. Taking a Cradle-to-Cradle approach means recycling is designed-in to the product lifecycle, rather than as an afterthought to a linear material journey. Reactive attempts at recycling must always fight the inertia of a system designed to optimise a one-way throughput of materials. Once materials are contained in a product, they are typically on a trajectory ending in landfill. Diverting those materials into new and modified processes of production is difficult. The Reee chair demonstrated that responsive recycling can be successful to a degree. The Embody chair and DESSO carpets represent a more progressive approach to designing not just products, but material systems. These examples seek to alter the trajectory of materials through the production and consumption process, by thinking ahead at the product design stage. Designing product systems on this principle is clearly a greater challenge than designing simply a chair or a carpet, but is required for a strategic use of materials.

Cradle-to-Cradle thinking is based on the principle of biomimicry – 'innovation inspired by nature' (Benyus, 2002). Biomimetic design looks to the natural world for potential solutions to human design problems. The surface structure of a lily pad forces water into droplets large enough to run off into a surrounding pond, preventing the pad from sinking under the weight of any water which falls on it. We have used this principle to create synthetic coatings for windows which make them self-cleaning, as dirt is washed away with rain droplets. To be more deeply biomimetic, these synthetic coatings would copy not only the functional principle of the surface of the lily pad, but would also be created according to natural principles using green chemistry. In a natural system, all waste becomes food for another part of the system. As one element of a system dies and decays it feeds the growth of another part. Human-made products, in contrast, are usually not recognised by natural systems, because they have been synthesised into forms not found in nature. As a result, human products typically become waste at the end of their life. We have created a new category of potentially polluting materials for which we have no use and natural systems have no appetite.

The Cradle-to-Cradle design protocol focuses on maintaining healthy and distinct material cycles. It seeks to design-out waste by identifying which of two metabolisms any material we use should enter at the end of a product's life. 'Biological nutrients' are products of nature, which can return to natural systems for reprocessing. 'Technical nutrients' are synthetic products of human manufacture, which must be kept in a technical metabolism managed by us (McDonough & Braungart, 2002). It may even be possible to improve the quality of human-made technical materials, via upcycling, if they are kept within their own closed systems (McDonough & Braungart, 2013). Pollution occurs when technical nutrients enter the biological metabolism. Natural systems cannot process or safely absorb human-made products such as oil-based plastics. The Great Pacific Garbage Patch refers to an ever-increasing mass of floating debris brought together by global ocean currents. There is no single 'garbage patch', but constantly moving concentrations and combinations of marine debris, much of it plastic, in the North Pacific Ocean (National Ocean Service, 2020). Plastic products will eventually disintegrate when exposed to the

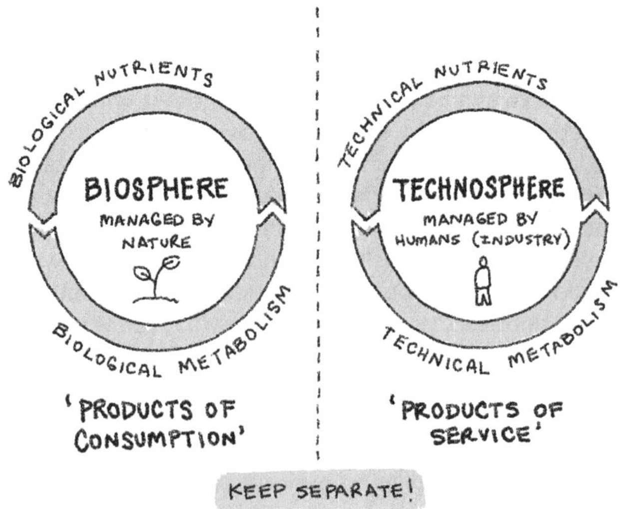

Figure 5.3 Cradle-to-Cradle design protocol.

forces of nature, but they only reduce into smaller pieces; they will not be accepted into nature's material cycles.

To avoid adding to the problem of material pollution, we have two options: (1) lock technical nutrients into closed and managed material cycles, ensuring they do not enter the biological metabolism; (2) replace them with biological alternatives that can safely enter the biological metabolism. This gives two design options: (i) redesign products and materials for effective recovery and recycling of technical nutrients; (ii) design-out the need for recycling by replacing technical nutrients with biological alternatives. The first approach requires material recovery systems that can reclaim oil-based waste plastics for reprocessing and reuse. This is difficult to do well. There are many kinds of plastics, and they are often locked into products that do not allow them to be easily separated. The fuel and energy costs of transporting and reprocessing lightweight but bulky plastic wastes can also be high. We also do not generally know what to do with the secondary plastic created, as our industrial system of production and consumption is not designed on circular principles. The second approach – replace technical nutrients with biological alternatives – substitutes bioplastics for oil-based plastics. These plastics made from plant-based materials biodegrade when no longer needed, ideally by composting in natural conditions (although some will only break down in specific industrially created environments). A technical metabolism is biomimetic – it seeks to mimic natural systems in the form of an artificial recycling infrastructure. Using biological nutrients within an existing biological metabolism is not imitating nature, it is using nature. This second approach works within, rather than outside of, natural systems. Technical nutrients and biological nutrients must be kept apart. It can, however, be difficult to keep technical and biological metabolisms separated. Biodegradable plastic bags are increasingly being introduced by retailers as a response to the problem of plastic waste. If we put them in the recycling collection for conventional plastic bags, biodegradable bags contaminate the waste stream, interfering with a technical metabolism. We need to collectively commit to one approach, or learn to better handle different materials which we currently fail to distinguish. Recycling is a behavioural, as much as a technical, issue.

There is a Cradle-to-Cradle certification, which assesses products using five categories: material health; material reutilisation; renewable energy and carbon management; water stewardship; social fairness. We see here that product sustainability is about more than material recycling. Products certified at the 'Basic' level meet minimum requirements, and show a commitment to improvement. 'Platinum' level certification is for a product designed and produced for a Cradle-to-Cradle world. Of the 400 product certifications issued at the time of writing, most were mid-level ('Silver'), including the Embody chair discussed above. None were yet Platinum, indicating that delivery of the Cradle-to-Cradle vision is some way off (Cradle-to-Cradle Products Innovation Institute, 2020). Embracing Cradle-to-Cradle thinking and switching to a circular economy requires us to 'remake the way we make things' (McDonough & Braungart, 2002). Cradle-to-Cradle is a call for system change, whereby we limit our use of materials to those which can be accommodated in either the biological metabolism of natural processes of decomposition, or an industrially controlled technical metabolism. Legislation such as WEEE, instrumental in the story of the Reee chair, is a required element of this shift. Another is materials pooling, by which partners share access to a common supply of a non-biological material, pooling information and purchasing power to create a viable 'open loop'. It is more efficient to do this collectively than individually. This requires some relaxation of the normal competitive principles of business, in which technical expertise is often protected rather than shared. This element of the Cradle-to-Cradle protocol is essential when dealing with high-value but low-volume technical materials such as synthetic fibres. A materials pool requires collectivism rather than competition, not just within a market sector (e.g. outdoor clothing) but ideally amongst all companies using a technical nutrient. An athletic shoe manufacturer, a furniture design firm and a high-tech materials company might all draw from a materials bank to create their new products, replenishing it with used products they recover and return for recycling. Cradle-to-Cradle contains a vision for building a community of businesses that share materials, information and values. The Cradle-to-Cradle brand would therefore unite collaborating brands from across sectors (McDonough & Braungart, 2002, p. 3).

Many companies now promote their own product take-back scheme. The technology company HP collects its own used printer cartridges from consumers to be reprocessed into new products, including cartridges. It supplements this supply of recovered material from its own products with plastic from recycled drinks bottles (HP, 2020). This represents a partial closed loop, and also a step towards the materials pooling advocated by Cradle-to-Cradle. The North Face is a globally leading producer of outdoor technical clothing. Its Clothes the Loop scheme invites customers to donate end-of-life garments and shoes in The North Face stores, no matter what the brand. Donated items are sorted into over 400 categories. They are then repurposed for reuse, or recycled into raw materials for use in products such as insulation, carpet padding, stuffing for toys or fibres for new clothing (The North Face, 2020). This is an open loop, in which the company accepts more than just its own products and contributes to numerous material pools (The North Face does not make toys, carpets or insulation). The openness within and between industries required by Cradle-to-Cradle is therefore increasingly in evidence.

Cradle-to-Cradle seeks to rescue materials from the lowest levels of the waste management hierarchy. Incineration is usually considered a last resort, preferable only to disposal in landfill. By burning waste to generate energy, we capture only its calorific value. A more considered use of incineration is, however, proposed by the Metablaze model (Banwell, 2015). Metablaze is a strategic response to the challenge of how to manage material cycles effectively. Presented as a pragmatic supplement to the Cradle-to-Cradle protocol, it seeks to replace reactive recycling with incineration, proposing we stop trying to recycle products that resist recycling, and use an alternative method to recover their material and calories. In addition to the biological and technical nutrients of the standard circular economy model, Metablaze seeks to capture material value in the ash produced by waste incineration. 'Digestible' materials are biological nutrients. 'Reusable' materials are technical nutrients, to be recovered from disassembled products and reused. 'Burnable' materials are what are left over, comprising technical nutrients which are too complex or degraded for recovery and reuse, and biological nutrients which are too contaminated to be reintroduced into the biological metabolism. The ash produced from burning these materials is carefully mined to recover the original constituent elements of the complex pre-burned material. Incineration is here analogous to the microbial action by which biological materials are broken down into their original elemental states. Burning returns resilient synthetic materials to their initial state, restoring them to reusable form.

Metablaze seems counter-intuitive. Incineration is generally seen as a crude way of capturing some residual value from waste by burning it to produce heat and then electricity. The ash is usually seen as a potentially toxic waste product. Metablaze identifies this ash as having significant value as a source of recoverable and reusable material, such as metals. If feasible, this seems a very pragmatic approach. Treating complex or problematic materials as burnable removes the need to redesign our material recovery systems. Sophisticated products such as fibreglass wind-power turbine blades are currently too challenging to recycle (Martin, 2020). They are designed for performance in service, but not end-of-life, and generate a waste stream that we then have to try to react to when the blades come out of service. Turbine blades are an example of the 'unmarketables' which Cradle-to-Cradle challenges us to stop using precisely because there is no long-term plan for

Figure 5.4 Metablaze icons for digestible, burnable and reusable materials.

their materials (McDonough & Braungart, 2002, p. 116). Metablaze instead proposes that we accept the existing trajectory of materials innovation towards ever-more complex high-performing composites. Rather than striving to change how we make products such as turbine blades, we should simply alter our approach to recovering value at the end of a product's useful life. Metablaze is presented as a pragmatic response to the difficulties faced by existing material recycling systems. It works with, rather than against, existing practices, and introduces incineration as a supplementary method of extracting technical nutrients from current waste streams. Burning of course produces emissions, and the 'burnable' material icon shows grams of CO_2 equivalent released per kilogramme of a particular material burned. Transparency over emissions is certainly needed generally, although often difficult to state with accuracy. In this case it would allow us to judge the merits of recovery from incineration more fully. Incineration with material recovery is, however, likely to be better than incineration without recovery.

NOVELTY OR NECESSITY

As discussed in Chapter 1, recycling and reuse have a long history. Reasons for doing them have, however, changed.

> The poor, the thrifty and the imaginative have always recycled. Clothing is worn and re-worn, remade and handed down, its scraps ultimately reused in quilts and toys. Food cans become prized materials. Wine bottles are remelted and transformed. Tires yield shoe soles. We begin our investigation with examples of these ingenious and often witty objects. (Arango Design Foundation, 1996)

This is the introduction to an exhibition catalogue celebrating 'Good everyday design from reused and recycled materials'. Material reuse can be considered 'witty' only if we do not need to make products this way. Recycled products can be ingenious without being necessary. An ashtray formed from a partially melted glass wine bottle is more of a novelty than a meaningful response to redeeming waste materials. This often seems particularly true of products which are explicitly promoted as being recycled, as if that compensates for their functional or aesthetic failings. For those who are materially poor, however, recycling and reuse are necessities, and products made in this way may be the only ones available. We should be cautious of assuming too much about an object of reuse and recycling until we know more about where and why, and not just how, it was made. A recycled object is always a product of a specific context, and what it represents will be informed by that context. Reuse of scavenged materials in developing economies should not be thought 'witty' – this is subsistence-based making. In developed Western economies, reuse and recycling are often motivated by environmentalism and concerns of sustainability. Recycling is likely to be ideological, and a choice rather than a necessity. We have other materials. In less materially rich economies, reuse of materials is likely to be driven by scarcity, however, and a greater need for self-reliance and self-sufficiency. The motivation of someone scavenging waste materials from a city waste dump in India is economic rather than ecological. Plastic bottles are collected to be sold, and so earn a basic livelihood. 'Waste' is a resource that is currently in the wrong place, and perhaps in the wrong form. Scavengers recover such resources and make them available to whoever recognises their value and will pay money for it. Informal recycling practices in countries such as India show how trash can be challenged as a concept when money can be made from recycling it.

The documentary *Landfill Harmonic* tells the story of a garbage picker, a music teacher and a group of children in Paraguay who out of necessity make their instruments out of garbage (Landfill Harmonic, n.d.). Coming together to form a self-styled 'Recycled Orchestra', these young people are able to access music making in a way which would not otherwise be possible. This is recycled design as necessity rather than novelty. Elsewhere, repurposing materials to make a musical instrument can have a different motivation. A cigar box guitar is a primitive version of the familiar instrument, simple in construction and relatively easy to learn how to play. Making your own cigar box guitar can be sold as an activity, with all materials, tools and tuition provided. In this case we pay to be guided through the process of making an instrument from basic and repurposed materials which we then play at the end of the class in another version of a recycled orchestra. This second example is really about the novelty of the experience, however, rather than any need to make an instrument in this way.

There are many such examples. We are privileged if we use corrugated metal sheets only for the walls of our garden compost heap, not our home. We are fortunate if we do not need to cut up old car tyres for rubber for makeshift shoes. We might use the caps of beer bottles as pieces in a board game, but we choose to do so. We pay a premium price for the beer bottles transformed in the UK into wine glasses, or a shopping bag made in the Philippines from locally sourced drinks cartons, discussed in Chapter 1. The geographies of these two products are different, in terms of who remakes what, for whom, where and why. Transforming UK waste for UK reconsumption at least keeps the recycling loop local. It makes less sense to buy the waste of other parts of the world, however its value is increased.

We have more incentive to recycle the more we clearly see its value. Even better when it responds directly to our needs. The Waste to Wealth programme focuses on the specific challenges faced by people living in urban slums in Africa. It seeks to generate income through recycling by encouraging entrepreneurialism involving community groups (Waste to Wealth, 2020). Products are made from local waste materials, for local use. Recycling tackles the waste problem and is a vehicle for delivery of social and economic development projects which address other challenges. Transforming plastic waste into moulded shoes and paving stones for immediate local use provides an enterprise model in which residents, local authorities and the private sector work together on development challenges. Local ownership improves the chances of these initiatives being sustained over time – a key principle of development economics. Aid programmes can create dependency, and projects often end when support is withdrawn. Ventures which are managed independently of outside involvement are more likely to endure. Waste to Wealth consciously links recycling activities to broader development agendas. This contrasts with recycling in developed economies, which too often produces inferior products with limited appeal. Recycled products offered to affluent consumers often appear to be in search of a clear purpose, and rarely find one beyond making us feel slightly better about our consumption. The fact that we can make a shopping bag from drinks cartons does not mean that we should, especially if we use it to buy things we could probably do without.

Recycling is more common than reuse in developed economies. We are more likely to wash a glass jar and put it in a recycling bin than we are to reuse it. We are collectively more comfortable with recycling than reuse as a cultural practice, perhaps because recycling allows us to avoid facing up to our responsibilities in relation to the problem of waste. Promotional campaigns encouraging us to recycle more always tell us how easy it is. The proliferation of collection bins for things we don't need any more means we don't need to think about what happens to them. An empty cola can, bought minutes before, can be thrown to that magical place – 'away'. This convenience prevents us mindfully engaging with our wastefulness. Recycling collection systems try to make correct disposal of products as easy as possible. By offering a way out of the dilemma of what to do with a piece of packaging that has served its short-term purpose, but retains its material value, they are complicit in our entrenched throwaway culture.

A challenge on the online open innovation platform OpenIDEO focused on sharing strategies for successful at-home waste sorting.

> Let's learn from 'expert recyclers' out there and find out why recycling is second-nature to some and not others. Perhaps you've lived in a different country and noticed a different attitude to waste? Are you familiar with a home recycling scheme that works really well and could easily be replicated in homes everywhere? Maybe it's a neighbour or friend or even a hotel or resort you've visited, that has an incredible recycling system in place. (OpenIDEO, n.d.)

The aim is to enable new recycling habits that could work in different countries and communities. The hope is for universal solutions that could be applied anywhere. This does ignore cultural factors, however, which may require some translation of a successful scheme from one context to another. The challenge also simply addresses the front-end of the recycling process – collection. The winning ideas therefore focus only on the first step of a material's recycling journey. The real challenge comes further along that journey, in effectively reprocessing and reusing waste collected for recycling and embodying it in successful new products.

Make Do and Mend was a leaflet issued by the British Ministry of Information during the Second World War. It contained thrifty design ideas and advice on reusing and repairing clothing. An updated version offered similar frugal advice for the 21st century (Norman, 2007). We would expect an emphasis on recycling and reuse to arise in response to a scarcity of resources, and to recede in times of plenty. For those who lived through wartime, however, thriftiness was not easy to abandon. Our present focus on recycling is not a direct response to immediate scarcity. It is our concerns about waste, not the potential exhaustion of reserves of available resources, which drives our recycling. 'Peak oil', for example, does not mark the point at which we begin to run out of the raw ingredient for plastics. It rather denotes the point of maximum extraction and demand. This peak has been anticipated many times, but has always been delayed by discovery of new reserves or advances in technology. The timing of the moment of peak oil will depend on which scenario we choose for our future energy supply, and is a point of our choosing (International Energy Agency, 2019). We can still obtain the oil we need to fuel our consumption. There is no need for thrift due to lack of supply; we are, however, turning away from use of oil due to our awareness of its contribution to Global Heating.

Collection of household waste intended for recycling rose in the UK from 11% in 2000 to 40% in 2010 (DEFRA, 2015, p. 32). Since then it increased less dramatically to 45% in 2018 (DEFRA, 2020, p. 4). The volume of waste generated by UK households held steady from 2015 to 2018 (DEFRA, 2020, p. 3). This suggests that we may have reached 'peak recycling' in the UK. Participation in recycling is not universal. There are regional variations between the countries of the UK

(DEFRA, 2020, p. 3). Recycling rates in developed economies also seem to be influenced by our level of income, where we live and a range of other social and demographic factors such as age and gender (Schultz *et al.*, 1995; Martin *et al.*, 2006; Langley, 2012). The imperative to recycle in economically developed nations such as the UK is not driven by individual or collective need. It does, however, rely on individual participation. If recycling has peaked, on this basis, due to a range of social and cultural factors, we have another reason to reconsider its role within a collective waste strategy.

GLOBAL RECYCLING

Cultures of recycling and reuse are different in developed and developing economies. The Western mass-recycling system is, however, global in its reach. Local collection of material for recycling is often followed by activity in other parts of the world. As in the example of e-waste processing, this can have extremely harmful social and environmental effects. As we saw in Chapter 4, in Ghana computers are crudely burnt in the open air to recover the copper in their cabling. The designer Hal Watts responded by developing a safer and more effective way to retrieve the copper. Computer cable is first mechanically ground into granules using a device mounted on the back of, and powered by, a bicycle. The copper and plastic granules are then separated via water flotation. The granulator and separator are both designed to be manufactured and maintained in Ghana. The designs are made available to local workshops who then produce and sell the machines to local recyclers. This seeks to create a locally based system which generates value for people in Ghana, as they more safely reprocess potentially toxic imported waste. The system is more efficient in recovering value from both elements of the cabling – copper is worth more unburnt than burnt, and the plastic insulation is retained rather than lost (Watts, 2012).

Social enterprise company Fairphone addresses dangerous material recovery practices in Ghana by relocating them to Europe, where they can be done more safely within a formal electronics recycling sector. Discarded phones are collected in Ghana and shipped to Belgium for efficient recycling, so the recovered materials can be used directly in new phones. This represents a holistic and responsible approach to product lifecycle management. Recycling e-waste is difficult and hazardous. Exporting it to parts of the world where labour can be more easily exploited and environmental damage done more readily is unsustainable in every sense of the word. Most mobile phones discarded in Europe still work. Many of those sent to Ghana therefore still have potential useful life. Fairphone works with local partners to first give a retrieved phone a second life in Ghana; only then is it brought back to Europe for recycling (Fairphone, 2017).

A smartphone is a complex product. As a result it is difficult to state simply where it was made. It contains minerals such as tantalum, tin, tungsten and gold, which are often mined in places affected by armed conflict and lack strong safety and environmental regulations. A component of a phone could traverse the globe before it is ready to be added to the product. Final assembly takes place in China. The materials in a phone have made an immense journey before the product reaches its point of sale in, for example, Europe. In travelling to Belgium for reprocessing, a Fairphone is not therefore returning to its place of manufacture, or remanufacture if its materials are to be used in a new product. This example demonstrates the challenge of implementing a strategic Cradle-to-Cradle approach to design for recycling. This must include the design not just of products and their processes of production, but also the systems by which the product lifecycle is managed to allow for multiple reincarnations of the materials used. US outdoor clothing company Patagonia was able to provide detailed analysis of the anticipated impacts of its Common Threads garment recovery and recycling programme when it was launched in 2005. Damaged and worn garments were to be returned for recycling into new products, ideally after repair options are exhausted. The company evaluated energy use and greenhouse gas emissions resulting from three scenarios: (1) making polyester from virgin materials; (2) making polyester using recycled garments that were collected locally in Japan, the site of the factory; (3) making polyester using Patagonia's own recycled garments, collected in the US. The analysis showed clear benefits to the recycling processes over non-recycling. The local Japan-based recycling process was best, as there was no need to transport materials to Japan from the US (Patagonia, 2005). These figures relate only to the production of the base textile material for garments. Other impacts, for example shipping products to markets worldwide, need to be added for all scenarios to get a fuller picture. To be viable, such a programme requires the participation of customers to donate their used products, and partnerships with other recycling organisations to provide the collection, sorting and reprocessing infrastructure. Patagonia has since made a wide range of additional reuse and repair options available to its customers, moving up the waste hierarchy from the initial focus on recycling in the Common Threads programme. Refurbished garments are sold on its website alongside brand-new products, accompanied by encouragement to choose old before new (Patagonia, 2020). This is one example of recycling being put in its proper place by a company that recognises more preferable alternative approaches to retaining the value of products, and successfully communicates this to its customers.

The Swiss coffee brand Nespresso has led a transformation in how affluent consumers drink coffee at home. A single portion of ground coffee is pre-packaged in a sealed aluminium and plastic capsule, through which hot water is forced by machine

SCENARIO	VIRGIN MATERIAL	LOCAL RECYCLING (JAPAN)	NON-LOCAL RECYCLING (USA)
ENERGY (MEGAJOULES)	72,000	12,000	18,000
CO2 EMISSIONS (TONNES)	4.2	1.0	1.2

Figure 5.5 Energy use and greenhouse gas emissions per tonne of base material, from three manufacturing scenarios

Source: Patagonia (2005)

to deliver an espresso. Nespresso capsules have become the industry standard for these machines and other manufacturers of capsules. Aluminium is used because it is the best material to protect the flavours and aromas of the coffee. These single-use capsules, however, constitute a significant new waste stream, as they are discarded once their contents have been used to make a cup of coffee. Aluminium is of course a high-value material which is also highly recyclable (see Chapter 2). Used coffee capsules therefore present a ripe opportunity for collection and recycling. Nespresso collects used capsules from consumers in three ways: (1) collection points in Nespresso stores; (2) other local collection points; (3) direct doorstep collection. The fact that capsules contain both aluminium and plastic means they have to be treated as an isolated waste stream, and cannot be combined with other materials. Owners of a capsule coffee machine require a steady supply of capsules to replenish it, and generate an equally steady stream of spent capsules. Consumers are therefore encouraged to join a Nespresso club membership programme which delivers new capsules by post. The company can thereby retain control of the supply and collection of full and empty capsules – ideal conditions for an effective closed loop recycling system. Nespresso states that the complexities of the recycling process, and varying logistics of capsule recovery in different countries, make a standardised approach to recycling difficult. It therefore participates in some existing national waste recovery schemes such as Green Dot (see Chapter 3), and sometimes develops its own. Nespresso supplies customers around the world. Standardised product design in the form of a universal capsule is not yet matched by standardised product recovery systems at end-of-life (Nespresso, 2020).

Nespresso has reached a global recycling rate of 30% (Nestlé-Nespresso, 2020). The company identifies four elements to increasing this recycling rate: (1) raising consumers' awareness and participation in recycling; (2) working with authorities to integrate capsule recycling into collective recycling schemes; (3) increasing the number of collection points; (4) integrating other portioned coffee manufacturers into its recycling scheme to improve availability and accessibility of aluminium capsule recycling. Nespresso is the leading producer of coffee capsules, but not the only one. Other companies make their own versions, all compatible with Nespresso machines. Nestlé, the owner of Nespresso, is therefore seeking an industry-wide response to the challenge of recovering and reusing aluminium from spent coffee capsules (Nestlé-Nespresso, 2020). This is materials pooling as advocated by the Cradle-to-Cradle design protocol, and has advantages over separate proprietary closed loop systems for different brands. This constitutes a whole-lifecycle approach to the use of aluminium in portioned coffee, from bauxite mining for aluminium ore through to post-consumer recycling of capsules. This is the beginnings of a community-building process in which competitors become partners in the recovery and reuse of a common resource via development of a shared circular supply of aluminium.

Nespresso's stated goal is to recycle all of its capsules, within a totally effective technical metabolism. It now includes recycled content in some new capsules, and its capsule machine parts (Nestlé-Nespresso, 2020). Other products made with recovered capsule aluminium – including a pen and a bicycle – have been made in partnership with other companies. A totally effective technical metabolism would see all new capsules made from recycled old capsules, with no loss of material or new extraction of material. Surmounting this technical challenge would for Nespresso constitute closure on the sustainability question. As with the Cradle-to-Cradle protocol itself, this, however, ignores the significant question of the energy, and its associated impacts, required to operate such a distributed system.

We could of course simply attempt to refill and reuse our spent capsules, or use the more durable refillable capsules that are now available (although not from Nespresso). For business reasons, Nespresso prefers to focus on recycling. Its response to the inherent sustainability challenge of repeat delivery of pre-portioned coffee is to optimise a capsule recovery system, to function within a circular aluminium industry. However efficiently this system is managed, it expresses the distributed production and consumption model which is a driver of global unsustainability. If we succeed in designing-out

Figure 5.6 Nespresso aluminium coffee capsule recycling process.

material waste via a perfect closed materials loop, the impacts of distributed delivery and collection systems in the service of recycling remain. Capsule coffee, like carbonated drinks, has an environmental impact which recycling cannot remedy. The convenience seems too difficult to give up, however, hence the turn to recycling. Nespresso's environmental strategy is one of managing harm. Nestlé has become the world's largest food and beverage company through the pursuit of growth and profit. Nespresso is primarily committed to protecting its commercial interests. Its coffee capsule system is designed to ensure the brand retains as much economic value as possible. Environmental considerations are useful as long as they serve the goals of the business, hence the neglect of refill and reuse.

LOCAL RECYCLING

Locally concentrated, small-scale materials recovery practices have advantages over the fractured and dispersed systems envisaged by the likes of Nespresso. Localism is, after all, a central principle of sustainability. The unsustainability of our current economic system is in large part a result of its ever-increasing lack of localism. A consumer product typically has many components, each made by a different manufacturer, perhaps in a different part of the world. Final assembly of the product may then take place somewhere else. Fairphone is making a pioneering attempt to identify and map all the suppliers contributing to its newest smartphone. This is a hugely complex task, and reveals many layers of sub-contracting which top tier companies (the brand which appears on the product) are rarely aware of (Fairphone, 2020). 'Who made this product?' is a difficult question to answer. A global approach to recycling perpetuates this system, as we move recovered and reprocessed, rather than new, materials across huge distances. This is far from the restorative and regenerative economy called for by Cradle-to-Cradle.

A fibershed is a limited geographical area from which we source materials, specifically textiles (Fibershed, 2020). A fibershed puts boundaries on our resource base. Abiding by these boundaries means favouring local over imported materials and responding creatively to the challenge of meeting our needs using only what is available locally. A fibershed is a form of bioregion, an area defined by a unique combination of ecological systems which then form the basis of a distinct social culture. The bioregion is the traditional historical identifier of place and space. Our sense of regional identity is often determined not by present political or administrative boundaries, but by older and more enduring topographical features which have deeper significance for those who live in a particular place. The bioregion offers a future for more sustainable and ecologically grounded material economies. Vernacular architecture uses local materials to respond to local needs. Textiles and clothing were traditionally produced locally, using local materials to make garments suited to local climate, conditions and customs. The idea of the fibershed identifies the problems of our current textile production and consumption, in terms of how and where things are made. A fibershed is a local system deriving its value and meaning from not being large-scale and globally distributed. In supporting small-scale production, fibersheds represent an alternative to the dominant present form of industrial capitalism, in which communities in developing economies work to provide the material needs of global populations. Production at massive scale is seen as the error. The fibershed model does not therefore seek to replace natural systems via Cradle-to-Cradle design, which mimics nature, but to work within those natural systems, respecting their limits and constraints. It seeks to replace global industrial systems with networks of small, resilient and flexible systems that complement ecological systems. Mass-consumerism is integrated with global production systems, and requires transportation of goods and materials from source to producer to consumer, and perhaps back again if we recycle. The notion of a fibershed challenges not just what that system produces, but the system itself. Recycling efforts which sit comfortably within the present distributed system, and operate on the same scale, do not challenge the inherent problems deriving from that scale. FREITAG, the Swiss manufacturer of bags made from truck tarpaulins, now makes clothing using only biodegradable textiles sourced in Europe, in effect using a fibershed (Freitag, 2017). This reduces material options. Cotton, the most commonly used textile for clothing, is not produced in Europe. FREITAG instead uses hemp, flax and modal, all of which have much lower cultivation impacts than cotton. FREITAG's clothes are also compostable at the end of their useful life, retaining their status as biological nutrients. Non-compostable buttons can be unscrewed for reuse. FREITAG therefore allows easy separation of materials by type, and diversion to different next destinations. Producing locally and on a small scale allows easier coordination of the processes involved, and reduces harmful environmental and social impacts. When operating within a fibershed we re-localise production and consumption, designing-out the built-in problems of operating globally.

Compostable clothing is an idea we need to get used to. We are more likely to wear pre-owned clothes. Vintage clothing stores make a virtue of the fact that their stock is not new. They may even call this recycling, particularly if they upgrade garments through repair or embellishment. There are clearly environmental benefits to not buying new, especially if redistribution takes place locally. Fashion is fickle, however, and longevity hard to guarantee. Decoupling material use from the rapid turnover of fast fashion is a key challenge for our consumer economy. There is little incentive to design for durability when the clothes we wear have a social lifespan of only a few months. Local and low-impact reworking of our clothes can, however, serve to embed their materials in a local economy. The challenge is to retain those materials in extended cycles of use, beyond the short seasons of the marketing industry and our own need for frequent social renewal through what we wear.

Reuse of material at the end of a product's life is best achieved by designing products at the outset with their end-of-life in mind, using single and simple materials where possible. Bio-Knit is a process that uses a single textile material to provide multiple functions. Different non-chemical treatments produce alternate soft and flexible, or firm and robust textures. In imitation of natural material processes, a single material can be configured in different ways with different properties. Multi-material products are the scourge of recycling processes. Too often these products are too challenging to do anything with at their end-of-life. Materials which can be manipulated using only heat and pressure, without altering their chemical composition, can more easily feed into closed loop material cycles. Bio-Knit shows how we can design synthetic materials as technical nutrients with the flexibility to be used and reused in multiple ways, in multiple applications, multiple times. We can reduce our materials palette, without compromising our ability to design high-performing products. Simplifying what we design with will certainly influence what, and even how, we design. Designers typically consider desired product function or performance, then investigate and specify suitable materials. What we want a product to do dictates what we

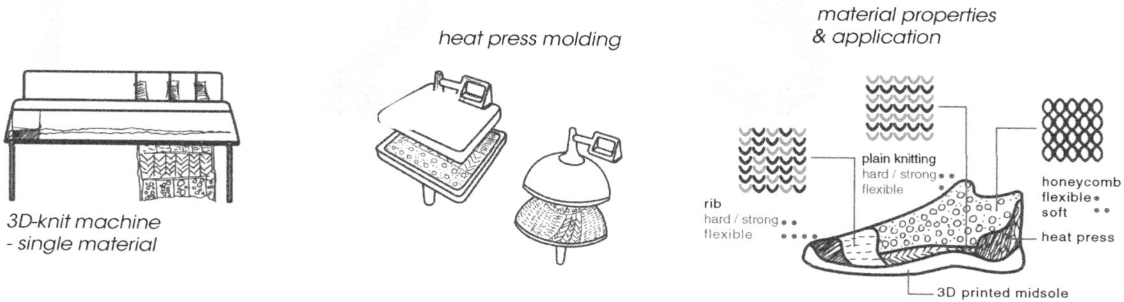

Figure 5.7 Bio-Knit multi-function textile.

make it from. Bio-Knit considers product design and material use equally. If a product can be made from one material, its different parts having different properties due to how they are manufactured, the recycling process is hugely simplified. By simplifying our use of materials, we may be able to manage their reuse more easily, without the need for complex distributed systems of recovery, reprocessing and resupply (Ammo Liao, n.d.).

The future of recycling must include new models of participation in which product users take active roles in giving materials new purpose. We currently donate unwanted materials to recovery processes in which we play no further part, unless we reacquire materials in new product forms by buying recycled products. We remain disengaged from the processes of re-production. There is scope for us to take a more active role in what happens to our unwanted materials. A new politics of recycling participation which puts users at the centre could yield greater engagement in recycling, and more useful outcomes via user involvement in the design of recycled products. Polyspolia is a proposal for a production system of domestic objects in which an item which breaks or falls out of favour can be easily unmade and remade in a new form (Yates-Johnson, n.d.). It combines 'poly' (many) with the Latin term ('spolia') for reuse of building material to create new monuments, to convey transmission of materials through successive generations of products. An object's materials are made visible to its user, and remain visible throughout its reincarnations via a product 'family tree'. A 'parent' object must be destroyed to provide material for its 'children'. If a parent is preserved, the family line ends. Yet 'childlessness' simply means that a product has reached its final ideal form, and has no need of reinvention or refinement. If a product no longer has value, we can seek to reconstitute its materials in a new object which will bear a family resemblance due the clear visibility of the inherited materials from which it is made. The model is currently demonstrated by simple solid forms, rather than useful products. This suggests the potential for user-led material recovery and product re-creation. For this potential to be realised, object users need to develop their materials literacy and design capability,

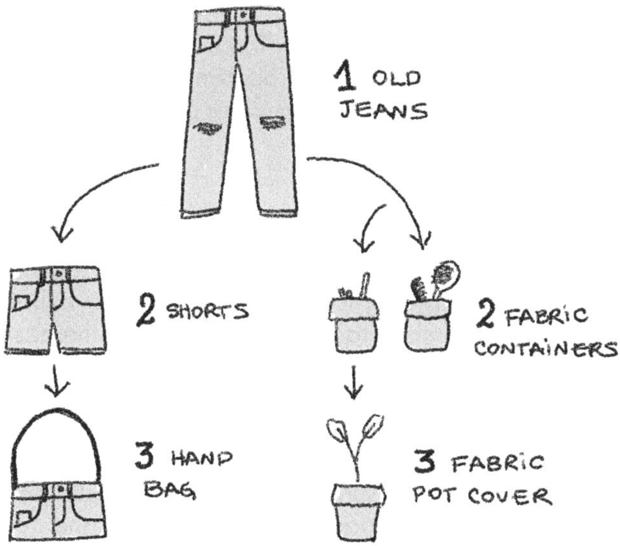

Figure 5.8 A product family tree showing generations of material reuse.

to become redesigners of their own products. This represents a potential democratisation of recycling, in which the user of the object to be recycled decides how, and into what, materials are remade. This possible future for recycling aligns with emerging practices of open design which give the user more involvement in what is made, where and how, on her behalf (van Abel *et al.*, 2011).

Polyspolia presents a vision for recycling which has many of the characteristics of reuse – it is done on a small scale and for direct personal benefit. It differs from reuse in that it involves transformation of materials, rather than simple repurposing of objects. It can be seen as a middle ground between reuse and recycling. People have recycled for as long as we have made objects using materials which outlive products. Industrialised Cradle-to-Cradle recycling systems are the latest technocratic manifestation of this. These sophisticated systems are, however, not applicable to much of the world as it currently is. Widely dispersed recycling infrastructures make little sense in terms of sustainability. Re-localising, reconnecting with our place in the face of the homogenising forces of globalisation, can be usefully applied to recycling. This can ensure that recycling relates directly to its context, and is not performed simply for its own sake.

CONCLUSION

This chapter considers future prospects for recycling as a worthwhile waste strategy, arguing for a greater focus on holistic design of recycling systems. Cradle-to-Cradle design, the basis of the idea of a circular economy, takes a strategic view of product and material lifecycles. It is, however, ecomodernist, and does not challenge the underlying unsustainability of our global system of production and consumption. A re-localisation of recycling is needed to ensure it can contribute to a more holistic and sustainable approach to how we make and consume things. Recycling has enjoyed unjustified popularity as a means of responding to the problem of our material waste. By putting it in context, both conceptually in relation to better options in the waste hierarchy, and geographically, we will make better use of recycling as we strive for a more sustainable future.

This chapter has taken a systems view of recycling, arguing for a more considered and strategic approach to what we make things from. This contrasts with the dominant current approach, which is typified by single use of materials and wishful thinking that we can deal reactively with the resultant waste. As consumers we feel trapped in behaviours for which waste seems the inevitable cost of convenience. The prospects for a more responsible use of recycling presented above all require us as consumers to play a more active part in material lifespans. The waste management hierarchy is a tool for all of us reduce, reuse, then recycle. Disruption of our current waste-making economies must come from all sides. To return to the earlier article with which this book shares its title:

> the current emphasis on individual behavioural change is unlikely to yield major changes in waste and other environmentally related practices unless

> such practices are also seen not only as simple issues of choice within a choice architecture framework ... but also a function of signals sent through the socio-economic and political system. (Barr *et al.*, 2013, p. 75)

It is clearly not solely down to us to make the change that is needed, but we cannot wait for the rest of the system to shift before we play our part. Recycling has long been a central pillar of environmentalism as a social movement. We must now move beyond recycling, and recognise its limitations as a route to a more sustainable future.

REFERENCES

Ammo Liao, n.d. Bio Knit [online]. Available from: https://www.ammoliao.com/bio-knit. [Accessed 1 September 2020].

Arango Design Foundation, 1996. *Re(f)use: Good everyday design from reused and recycled materials,* The Ninth Arango International Design Exhibition. Miami: Arango Design Foundation.

Banwell, E., 2015. Metablaze: Why Nature says we should be burning the sacred cows of cradle-to-cradle and how we can. Thesis (MA). Royal College of Art.

Barr, S., *et al.*, 2013. Beyond recycling: An integrated approach for understanding municipal waste management. *Applied Geography*, 39, 67–77. https://doi.org/10.1016/j.apgeog.2012.11.006. [Accessed 1 September 2020].

BBM Sustainable Design, n.d. Waste House [online]. Available from: https://bbm-architects.co.uk/portfolio/waste-house/. [Accessed 1 September 2020].

Benyus, J., 2002. *Biomimicry: Innovation inspired by nature*. New York: Harper Perennial.

Cradle to Cradle Products Innovation Institute, 2020. What is Cradle to Cradle Certified? [online] Available from: https://www.c2ccertified.org/get-certified/product-certification. [Accessed 1 September 2020].

Department for Environment, Food and Rural Affairs (DEFRA), 2015. Digest of waste and resource statistics – 2015 *edition* [online]. Available from: https://www.gov.uk/government/statistics/digest-of-waste-and-resource-statistics-2015-edition. [Accessed 1 September 2020].

Department for Environment, Food and Rural Affairs (DEFRA), 2020. UK statistics on waste [online]. Available from: http://www.defra.gov.uk/environment/waste/. [Accessed 1 September 2020].

Duvaltext, n.d. Four industry leaders join forces to reduce their environmental impact [online]. Available from: https://www.victortextiles.com/en/at-a-glance/#case-study. [Accessed 1 September 2020].

Fairphone, 2017. Collecting used phones, from Africa to Europe [online]. Available from: https://www.fairphone.com/en/2017/08/01/collecting-used-phones-from-africa-to-europe/. [Accessed 1 September 2020].

Fairphone, 2020. Mapping the journey of your Fairphone [online]. Available from: https://www.fairphone.com/en/impact/source-map-transparency/. [Accessed 1 September 2020].

Fibershed, 2020. About [online]. Available from: https://fibershed.org/about/. [Accessed 1 September 2020].

Freitag, D., 2017. From fibers to F-ABRIC. Available from: https://www.freitag.ch/en/fiberstofabric. [Accessed 1 September 2020].

Herman Miller, 2014. *Environmental product summary – Embody chair*. Zeeland, MI: Herman Miller.

Herman Miller, 2020. Design for the environment [online]. Available from: https://www.hermanmiller.com/our-values/environmental-advocacy/design-for-the-environment/. [Accessed 1 September 2020].

HP, 2020. Product return & recycling [online]. Available from: https://www8.hp.com/uk/en/hp-information/environment/product-recycling.html. [Accessed 1 September 2020].

International Energy Agency, 2019. *World energy outlook 2019*. Paris: International Energy Agency [online]. Available from: https://www.iea.org/reports/world-energy-outlook-2019. [Accessed 1 September 2020].

Landfill Harmonic, n.d. *Landfill Harmonic* [online]. Available from: http://www.landfillharmonicmovie.com. [Accessed 1 September 2020].

Langley, J., 2012. Is green a grey area? Sustainability and inclusivity: Recycling and the ageing population. *The Design Journal*, 15(1), 33–56.

McDonough, W., and Braungart, M., 2002. *Cradle to Cradle: Remaking the way we make things*. New York: North Point Press.

McDonough, W., and Braungart, M., 2013. *The Upcycle: Beyond sustainability – designing for abundance*. New York: Macmillan USA.

Martin, C., 2020. Wind turbine blades can't be recycled, so they're piling up in landfills [online]. Available from: https://www.bloomberg.com/news/features/2020-02-05/wind-turbine-blades-can-t-be-recycled-so-they-re-piling-up-in-landfills. [Accessed 1 September 2020].

Martin, M., Williams, I. and Clark, M., 2006. Social, cultural and structural influences on household waste recycling: A case study. *Resources, Conservation and Recycling*, 48, 357–395.

National Ocean Service, 2020. What is the Great Pacific Garbage Patch? [online]. Available from: https://oceanservice.noaa.gov/facts/garbagepatch.html. [Accessed 1 September 2020].

Nespresso, 2020. Recycling [online]. Available from: https://www.nespresso.com/uk/en/recycling. [Accessed 1 September 2020].

Nestlé-Nespresso, 2020. Nespresso takes important step towards circularity with launch of new capsules using 80% recycled aluminium [online]. Available from: https://www.nestle-nespresso.com/media/mediareleases/nespresso-launches-capsules-using-80-recycled-aluminium. [Accessed 1 September 2020].

Norman, J., 2007. *Make do and mend: Keeping family and home afloat on war rations*. London: Michael O'Mara Books.

The North Face, 2020. Clothes the Loop [online]. Available from: https://www.thenorthface.co.uk/innovation/sustainability/product/clothes-the-loop.html. [Accessed 1 September 2020].

OpenIDEO, n.d. How might we establish better recycling habits at home? [online]. Available from: https://challenges.openideo.com/challenge/recycle-challenge/brief.html. [Accessed 1 September 2020].

Patagonia, 2005. Patagonia's Common Threads garment recycling program: A detailed analysis [online]. Available from: https://www.patagonia.com/on/demandware.static/Sites-patagonia-us-Site/Library-Sites-PatagoniaShared/en_US/PDF-US/common_threads_whitepaper.pdf. [Accessed 1 September 2020].

Patagonia, 2020. Worn Wear [online]. Available from: https://wornwear.patagonia.com. [Accessed 1 September 2020].

Pli Design, 2008. Press release [online]. Available from: http://www.plidesign.co.uk/news/2008/08/press-release-product-launch-of-first-uk-made-chair-to-use-100-recycled-plastic-from-a-single-waste-source. [Accessed 1 September 2020].

Schultz, P.W., Oskamp, S. and Mainieri, T., 1995. Who recycles and when? A review of personal and situational factors. *Journal of Environmental Psychology*, 15, 105–121.

Stuart, K., 2013. PlayStation 2 manufacture ends after 12 years [online]. Available from: https://www.theguardian.com/technology/2013/jan/04/playstation-2-manufacture-ends-years. [Accessed 1 September 2020].

Tarkett, 2020. DESSO Carpet tile and roll solutions: Innovation, functionality and *sustainability* [online]. Available from: https://professionals.tarkett.com/en_EU/node/desso-carpet-tile-and-roll-solutions-innovation-functionality-and-sustainability-5527. [Accessed 1 September 2020].

van Abel, B., Evers, L., Klaassen, R. and Troxler, P., 2011. *Open design now: Why design cannot remain exclusive*. Amsterdam: BIS.
Waste to Wealth, 2020. Waste to Wealth [online]. Available online: http://wastetowealth.livingearth.org.uk/. [Accessed 1 September 2020].
Watts, H., 2012. Esource [online]. Available from: https://cargocollective.com/halwatts/Esource. [Accessed 1 September 2020].
Yates-Johnson, Will, n.d. Polyspolia [online]. Available from: http://www.whyj.uk/polyspolia. [Accessed 1 September 2020].

INDEX

Note: Illustrations are denoted in *italics*.

For Product Safety Concerns and Information please contact our EU representative GPSR@taylorandfrancis.com Taylor & Francis Verlag GmbH, Kaufingerstraße 24, 80331 München, Germany

Printed and bound by CPI Group (UK) Ltd, Croydon, CR0 4YY
14/09/2026
02216240-0001